RAND NATIONAL DEFENSE RESEARCH INSTITUTE

Approaches to Strategic Sealift Readiness

Bradley Martin, Roland J. Yardley

Prepared for the United States Navy

For more information on this publication, visit www.rand.org/t/RR3049

Library of Congress Cataloging-in-Publication Data is available for this publication.
ISBN: 978-1-9774-0276-9

Published by the RAND Corporation, Santa Monica, Calif.

Cover: USNS Mendonca offloads cargo in Virginia Beach, Va. (U.S. Navy photo).

Preface

To fight in remote areas, the U.S. military must be able to move large amounts of cargo as required by the operational plans of combatant commanders. Strategic sealift—a fleet of 61 commercial-standard ships—plays a central role in meeting these transportation requirements. Given the importance of the strategic sealift fleet, the U.S. Navy is interested in ensuring that these ships are ready to respond when the need arises. The fleet is managed by two organizations, the Military Sealift Command, the naval component of the U.S. Transportation Command, and the Maritime Administration of the U.S. Department of Transportation. The Navy wanted a better understanding of how the different management models used by these organizations affected the readiness of the fleet. But the research quickly pointed to the fact that other factors—including how requirements are determined, material readiness, and personnel readiness—affected strategic sealift readiness as much as or more than organizational management. As a result, this report touches on all these topics as they collectively pertain to readiness of the strategic sealift fleet and offers recommendations on how readiness can be improved.

The findings and conclusions presented in this report will be of interest both to U.S. Transportation Command as it determines readiness for meeting operational requirements and to the Office of the Chief of Naval Operations in determining resource and investment priorities to meet these requirements.

This research was sponsored by the U.S. Navy and conducted within the Navy and Marine Forces Center of the RAND Corporation National Defense Research Institute, a federally funded research and development center sponsored by the Office of the Secretary of Defense, the Joint Staff, the Unified Combatant Commands, the Navy, the Marine Corps, the defense agencies, and the defense Intelligence Community.

For more information on the RAND Navy and Marine Forces Center, see www.rand.org/nsrd/ndri/centers/navy-and-marine-forces or contact the director (contact information is provided on the webpage).

Comments or questions concerning this draft report should be addressed to the principal investigators, Bradley Martin at bmartin@rand.org or Roland Yardley at yardley@rand.org.

Contents

Figures and Tables

Figures

Tables

Summary

The U.S. military must be able to move large amounts of military cargo on time lines dictated by the operational plans of combatant commanders when fighting in areas far removed from U.S. territory. To meet these transportation requirements when the need arises, the U.S. Navy maintains a fleet of 61 commercial-standard ships—referred to as the strategic sealift fleet. This fleet must be maintained to a certain level of readiness to respond when the need arises, and the Office of the Chief of Naval Operations (OPNAV) was interested in whether the readiness targets for the fleet are being achieved and how the management of this fleet—which is divided between the Military Sealift Command (MSC) and the Maritime Administration (MARAD)—affects readiness. Though organizational management plays a role, the findings of our research point to many other factors that also have a substantial effect on strategic sealift readiness—including requirements determination, material readiness, and personnel readiness. The research team concluded that each of these areas can be improved in ways that could collectively increase strategic sealift readiness and makes recommendations toward that end.

The Strategic Sealift Fleet and Readiness Goals

Strategic sealift is maintained by two different organizations under different readiness management constructs. The surge sealift fleet of 15 ships is maintained by MSC, the naval component of U.S. Transportation Command (USTRANSCOM), which uses the Transportation Working Capital Fund construct to maintain and operate these ships—a construct that requires users to pay the expenses associated with use of these ships. The Ready Reserve Force (RRF) of 46 ships is maintained by MARAD, part of the U.S. Department of Transportation. MARAD does not use a working capital fund construct[1] but instead relies on funding provided by the U.S. Navy as part of the annual appropriations and apportioning process.

1 Working Capital Fund (WCF) is a revolving fund, an account or fund that relies on sales revenue rather than direct congressional appropriations to finance its operations. It is intended to generate adequate revenue to cover

For purposes of sealift, the required readiness outputs are (1) the ability to meet real-world tasking on short notice and (2) readiness to support operational plan time lines for deploying forces and material. The resources required for ships to meet the required operational output, which we examine in this report, include

- scheduled maintenance availabilities in industrial facilities
- unscheduled repairs requiring specialized parts for installed equipment
- qualified personnel, specifically merchant mariners for reduced operating status (ROS) to conduct organizational maintenance and full operational status (FOS) crews to perform operational missions
- training to ensure crews are current in mariner skills
- operating days to support training and test equipment.

The ships in both fleets are held to the same readiness standard, which is that 85 percent of ships in the sealift fleet will be available within five days of activation. Although these two fleets are held to the same standard, they report different readiness levels. MSC has reported over the last year that approximately 71 percent of its fleet is ready to meet the five-day readiness standard,[2] while MARAD reports a higher number, an average of about 85 percent.[3] The difference in readiness reporting between the two organizations is among the reasons the OPNAV staff, as the resources sponsor for strategic sealift, requested a detailed look at practices for sustaining the sealift fleet.

Research Objectives and Approach

In responding to this request, we addressed six specific questions that apply to sealift readiness requirements and the mechanisms for generating this readiness:

- Is the sealift fleet ready to execute National Defense Strategy assigned missions?
- Are there sufficient ready ships and crews?
- How long would it take to reach the achieved readiness?
- What is the gap between the requirement and the provided capability?
- Do the organizational approaches yield different results?
- What is the relative cost of each approach?

---

the full costs of its operations and to finance the fund's continuing operations without fiscal year limitation. A revolving fund is intended to operate on a break-even basis over time; that is, it neither makes a profit nor incurs a loss.

[2] MSC email reports of readiness, March–July 2018.

[3] MARAD email reports of readiness, March–July 2018.

To conduct this analysis, we used a mix of data reported in various systems and the assessments of subject matter experts. In general, we preferred objective measurements, but we encountered cases where measurements were not available or were insufficient. For such cases, we augmented objective measurement with observations provided by subject matter experts, in particular those participating in and directly observing the processes.

Findings and Recommendations

While we used the above questions as our guidelines for research, our findings did not in every case conform exactly to the format we originally specified. By looking at limited-duration activations, we were able to establish that the sealift fleet can carry out those missions, but we found reasons to doubt that the force is postured for a larger-scale activation. At a minimum, the systems intended to demonstrate readiness were found to be ineffective in showing the actual readiness of the force required to meet larger-scale activations. The ineffectiveness of this system also brings into question any assessment of time required or the seriousness of the gap. While we found that this lack of reliability is present in both the RRF and the surge sealift force, we did find that the divided management construct had an impact on readiness generation. The following findings and recommendations describe our general findings more specifically.

The overall results of our assessment suggest that the readiness reporting systems in place today may not be fully accurate, missing nuances that could mean lower readiness in actual operational circumstances than is being reported. Aspects of the processes currently in place should be modified to provide a more accurate picture of strategic sealift readiness and more accurate requirements against which to measure. In addition, critical inputs to operational readiness—including material and personnel readiness—appear to be overlooked in understanding the overall readiness picture. Within this context, we offer recommendations in six areas: operational requirements, turbo activation (TA) practices, required operational capability, material readiness, personnel readiness, and management structure.

Operational Requirements

Operational requirements need to be stated clearly and to realistically reflect all constraint in the delivery system. Current sealift readiness requirements for arrival times are sooner than what the rest of the delivery system can reasonably support. Strategic sealift should not be required to be readier than the rest of the system.

- **Recommendations:** Formally revise the readiness requirement for sealift. Align readiness requirements to deployment needs as specified in operational plans, and realistically account for potential delays from other components.

Turbo Activation Practices

USTRANSCOM tests the readiness of the fleet by conducting a test of a vessel's ability to activate, through a no-notice TA process. When a vessel is directed to conduct a TA, all aspects of a vessel's activation must be completed within the five-day readiness period. While the number of vessels in the strategic sealift fleet has been relatively constant, the number of TAs has varied over time. Some ships have done several TAs while others have had few or none. Moreover, TAs do not accurately reflect what a vessel and crew would need to do to accomplish their mission.

Instead, to support a mission, a strategic sealift vessel should be activated, complete all necessary requirements to deploy, get underway, transit to the port of embarkation, and fully utilize all onboard equipment required to execute an actual mission. This type of activation test would provide many benefits, including a full test and run time for the vessel's engineering plant, experience for the crew, and a test of the full range of the platform's capabilities for which it is designed to operate. The trade-off is the cost in time and money, which competes with other resource needs.

- **Recommendations:** Revise the TA practice to regular activation of multiple units for multiple days underway to align with missions. Ensure that all ships receive a TA within a time frame to ensure operational readiness of that particular unit (and not as a representative test that the system generates ready units).

Required Operational Capability

Ships in the sealift fleet have an approved set of operational capability requirements that is intended to specify the kinds of missions these ships are required to conduct and the environment in which these missions are likely to take place. Some capabilities specified are outdated for strategic sealift vessels. Nevertheless, equipment needed to meet these requirements must be maintained to readiness standards even though it is no longer in use. Money, time, and resources are wasted. Although we recognize that updating standards is a time-consuming process, the savings in maintenance requirements could be substantial.

- **Recommendation:** Review required operational capability / projected operational environment statements for ships to ensure relevant requirements.

Material Readiness

A major unknown is the material condition of the many ships that have not recently activated. Even among those activated, the TAs have not been universally successful and for some ship classes notably unsuccessful. Sealift vessels must complete certifications as a condition for operations. Yet only a few ships are completely current with their certifications; most have several certifications that are not current, and many ships in each fleet were all missing required certifications. Expired certification does suggest that issues with the management of sealift ships are not being captured in TA performance or elsewhere. Investments are needed in many areas of material readiness to better ensure ships are ready and capable of executing their missions—ranging from more directly addressing the issue of parts obsolescence to common databases to improved oversight of repair contractors.

- **Recommendations:** Improve cost-benefit analysis for repair versus replacement of equipment. Improve documentation of deferred maintenance and, in particular, the mission impact of deferral. Use common access databases across the whole sealift enterprise. Streamline processes for alterations to replace obsolete and difficult to maintain equipment; obsolescent parts add measurable cost and seriously affect readiness; obsolescent plants are increasingly difficult to man and maintain. Improve oversight of ship management companies and repair contractors. Use a common standard for report of mission-limiting casualties. Invest in home port facilities—in particular, storm-protected harbor berths.

Personnel Readiness

Strategic sealift vessels are manned and maintained by a reduced number of crewmembers, called a ROS crew; when activated, the ROS crew is augmented by a FOS crew. ROS crews' responsibilities are much more focused on maintenance and sustainment of the vessel than FOS crews. ROS crew mariners with required licenses and all required training are difficult to find, and it is challenging for these crews to gain all certification and training demands as compared with FOS crews. Reduced and limited underway time play a big role. In addition, the pay for ROS crews is lower than FOS crews.

Navy leadership report that crew continuity is important to readiness and is a perishable resource, and the only way to pass knowledge on is to get underway with sufficient time at sea so that crews can obtain their certifications. There is a national shortage of qualified personnel that directly affects the ability to man the strategic sealift fleet. Mariner manning may be sufficient for initial activation, but activation crews may have long waits for a replacement crew, especially for steam engineers.

- **Recommendations:** Improve stability and capability of sealift crews. Conduct more frequent and longer underway periods. Review compensation packages.

Management Structure

The dual management structure of the surge sealift fleet does not appear to have originated from a clear decision with clear justification. Maintaining two management structures results in different reporting methods and maintenance tracking systems, among other differences. MARAD and MSC essentially accomplish the same missions with the strategic sealift fleet. MSC is the naval component of USTRANSCOM and has a variety of operational responsibilities, including operational oversight and control of activated sealift. It is in a better position to appreciate their core competency of military need than MARAD. MARAD by mission is best able to state how much and what kind of capability is needed and at what point. Realigning responsibilities along these lines would result in a more efficient and effective management structure. A management structure that assigns responsibility for military operational command and control to MSC and for the manning, training, and equipping functions to MARAD aligns more closely with each command's core competencies, reduces duplicate responsibilities for the management and maintenance of similar vessels to one provider (MARAD), and standardizes readiness reporting, as MARAD would be responsible for reporting on the entire fleet.

- **Recommendations:** Refocus MSC attention to USTRANSCOM component issues and away from day-to-day management of equipment and personnel. Assign the man, train, maintain, and equip functions of MSC vessels to MARAD, consistent with the U.S. Code Title 10 responsibilities of the OPNAV.

Concluding Thoughts

These recommendations are in some cases applicable regardless of management structure. More frequent and more varied TAs, more underway time for FOS crews, better documentation of material condition, and improved requirements review processes could all be implemented with or without a common management structure. However, the divided management structure has no clear justification or purpose and is in some respects detrimental to improvement. While changing the management structure to focus MSC on operational employment and MARAD on readiness generation does not really involve adding or subtracting staff; it would involve organizational change. Organizational change is often disruptive and should not be undertaken without good reason. The concerns surrounding strategic sealift's overall readiness appear to provide good justification.

Acknowledgments

We'd like to acknowledge the contributions of Avi Bergman and Nate Edenfield of the RAND Corporation research staff, as well as Barbara Bicksler for her expert help in preparing the manuscript. We'd in particular like to thank the sponsor, the Office of the Chief of Naval Operations Strategic Sealift Directorate, as well as stakeholders in the U.S. Transportation Command, Military Sealift Command, and the Department of Transportation Maritime Administration.

Abbreviations

ABS	American Bureau of Shipping
CPX	command post exercise
FOS	full operational status
MARAD	Maritime Administration
MSC	Military Sealift Command
OPNAV	Office of the Chief of Naval Operations
POE	projected operational environment
PREPO	pre-positioning
PSIX	Port State Information Exchange
ROC	required operational capability
ROS	reduced operating status
RRF	Ready Reserve Force
SAMM	Shipboard Automated Maintenance Module
STCW	Standards of Training and Certification for Watchstanders
TA	turbo activation
TPFDD	time-phased force deployment data
TWCF	Transportation Working Capital Fund
USCG	U.S. Coast Guard
USTRANSCOM	U.S. Transportation Command

CHAPTER ONE

Introduction

Background

For decades, the U.S. National Security Strategy has presupposed the need to potentially fight in areas far removed from U.S. territory, both to defend allies and to project power into unstable and contested areas. To conduct such operations, the U.S. military must be able to move large amounts of military cargo on time lines required by the operational plans of combatant commanders, as specified in movement plans generated by the U.S. Transportation Command (USTRANSCOM). For this purpose, the U.S. Navy has procured and maintained a fleet of 61 commercial-standard ships—referred to as the strategic sealift fleet. As we will discuss in more detail later in this report, while there is no stated readiness requirement for the whole strategic sealift fleet, each of the ships is expected to be ready for unrestricted underway mission operations within five days of being activated. Yet readiness reporting is not consistent for different parts of the fleet; why these differences exist is the central topic of this report.

Readiness Inputs and Outputs

All readiness models include a desired end state (the output) and a set of readiness inputs. For purposes of sealift, the required readiness outputs are (1) ability to meet real-world tasking at short notice and (2) readiness to support operations plans' time-phased force deployment data (TPFDD) time lines. These outputs are service operational requirements of USTRANSCOM, which uses a turbo activation (TA) system to test the ability of individual ships to meet required operational time lines.[1] The informally stated but not codified requirement for readiness is 85 percent of ships in the sealift fleet available for mission tasking within five days of being activated.

On the other side of the model, readiness inputs refer to the resources necessary for ships to meet the required operational outputs. These are funded by the U.S. Navy, although there are a variety of resource providers. These resources include

- scheduled maintenance availabilities in industrial facilities
- unscheduled repairs requiring specialized parts for installed equipment

[1] USTRANSCOM, Memorandum, May 19, 2015.

- qualified personnel, specifically merchant mariners for reduced operating status (ROS) to conduct organizational maintenance and full operational status (FOS) crews to perform operational missions
- training to ensure crews are current in mariner skills
- operating days to support training and test equipment.

It is difficult to determine the specific resources required for a vessel to reach a specific readiness level. The difficulty stems from the fact that resource areas intertwine, and adding resources in one area does not necessarily lead to improved overall readiness without comparable investment in related areas. For example, a poorly trained crew will have a difficult time performing effective maintenance or identifying abnormal conditions. Adding resources for depot maintenance would not necessarily result in better readiness outcomes unless sufficient investment is also made in personnel and training.

Different Readiness Constructs

Strategic sealift is maintained by two different organizations under different readiness management constructs. The surge sealift fleet of 15 ships is maintained by the Military Sealift Command (MSC), the naval component of USTRANSCOM, which uses the Transportation Working Capital Fund (TWCF) construct to maintain and operate these ships—a construct that requires users to pay the expenses. The Ready Reserve Force (RRF) of 46 ships is maintained by the Maritime Administration (MARAD) of the U.S. Department of Transportation. Consistent with MARAD's mission to promote a U.S. merchant marine capable of supporting U.S. national defense requirements,[2] these ships are held to the same standard as the surge sealift fleet: five-day ready for mission tasking at sea upon being activated. MARAD does not use a TWCF construct, instead relying on funding provided by the U.S. Navy as part of the annual appropriations process.

The reasons for the divided management construct are not completely clear. They are both held to the same readiness standard, draw from the same pool of repair and operational resources, and are subject to activation in the same time line. Both would be required for any large-scale activation, and neither is preferentially selected for more limited real-world missions.

Although these two fleets are held to the same standard, they report different readiness levels. MSC has reported over the last year that approximately 71 percent of its fleet is ready to meet the five-day readiness standard,[3] while MARAD reports a higher number, an average of about 85 percent.[4] We will discuss in greater detail the

---

[2] U.S. Department of Transportation, Maritime Administration, homepage, n.d.

[3] MSC email reports of readiness, March–July 2018.

[4] MSC email reports of readiness, March–July 2018.

potential reasons for these differences in reporting, but the marked difference is among the reasons the Office of the Chief of Naval Operations (OPNAV) staff, as resource sponsor for strategic sealift, requested a detailed look at practices for sustaining the sealift fleet.

Research Objective

In this report, we address six specific questions that apply to sealift readiness requirements and the mechanisms for generating this readiness:

- Is the sealift fleet ready to execute National Defense Strategy assigned missions?
- Are there sufficient ready ships and crews?
- How long would it take to reach the achieved readiness?
- What is the gap between the requirement and the provided capability?
- Do the organizational approaches yield different results?
- What is the relative cost of each approach?

The answers to these questions will be of interest both to USTRANSCOM for determining readiness for meeting operational requirements and to the OPNAV for determining resource and investment priorities to meet these requirements.

Approach and Organization of This Report

To conduct this analysis, we used a mix of data reported in various systems and the assessments of subject matter experts. In general, we preferred objective measurements, but we encountered cases where measurements were not available or were insufficient. For such cases, we augmented objective measurement with observations provided by subject matter experts, in particular those participating in and directly observing the processes.

Although one of the major reasons for undertaking this study is to find the best management approach for this mission area, we did not begin with the assumption that any management approach is best or most appropriate. Nor did we assume that determining a best management approach would resolve the readiness challenges facing the strategic sealift fleet because of the possibility that other systemic issues may exist that go beyond fleet management.

Accordingly, our first set of tasks related to defining the extent to which the whole fleet meets the requirements established in plan time lines (Chapter Two). Since the details of these plans are classified, we will discuss the level of readiness in general terms. For this, we relied on program-of-record readiness reporting systems, with minimal expert opinion required.

We then examined the reasons for the discrepancies in achieving readiness, which required an examination of the readiness inputs discussed earlier in this chapter. We broadly divided these inputs into material (Chapter Three), to include maintenance and supply support, and personnel (Chapter Four), to include manning levels and proficiency. Evaluating these areas required both collection of objective data and interviews with subject matter experts.

Finally, although we did not begin with a presumption that one management system is superior to another, we did identify management best practices that promote the best use of readiness inputs (Chapter Five). We codified these and use these as a basis for recommending organizational changes. These will be included as conclusions and recommendations (Chapter Six).

CHAPTER TWO

Readiness Requirement

Operational requirements are defined as capabilities commanders need to achieve objectives. The requirement might be put forth in the absence of resource constraints: that is, the command asking for the capability is making the request based on need rather than availability. By this measurement, not every requirement will necessarily be filled if resources are not available. Trade-offs between different requirements and available resources occur regularly in discussions between commands requiring capabilities and the military services or organizations providing them.

Strategic sealift as a capability is required by more than one combatant commander; most existing operational plans and most contingency operations require some level of transportation by sealift. Sealift is provided in a common user pool managed by the USTRANSCOM. USTRANSCOM manages immediate operational need and plans for major operations. Immediate operational need could conceivably be quite large, but, as we will discuss below, generally these requirements have been limited. Major operations involve larger commitments. In the remainder of this chapter, we examine the readiness requirements to support these operations.

Immediate Operational Need

Sealift has proven to be very useful as a tool for moving large amounts of cargo, ranging from Army brigade combat teams to Marine Corps aviation support to material for humanitarian assistance. This requirement is not fixed, being dependent entirely on real-world events. USTRANSCOM manages requirements through a TWCF, which means that users are required to pay the expenses, including operational and personnel costs. This fee for service creates a natural means for commands to express exactly how much they value a mission: Are they willing to pay the fee?

Between 2005 and 2017, operational requirements for both MSC and MARAD declined overall, as shown in Figure 2.1. While the size of the sealift fleet has not changed appreciably during this period, the number of days commands used sealift has declined from a peak in 2005 and fluctuated since then. Neither a lack of capacity nor a lack of readiness to support the mission appears to be an underlying cause for this

Figure 2.1
Operational Days for Military Sealift Command and Maritime Administration

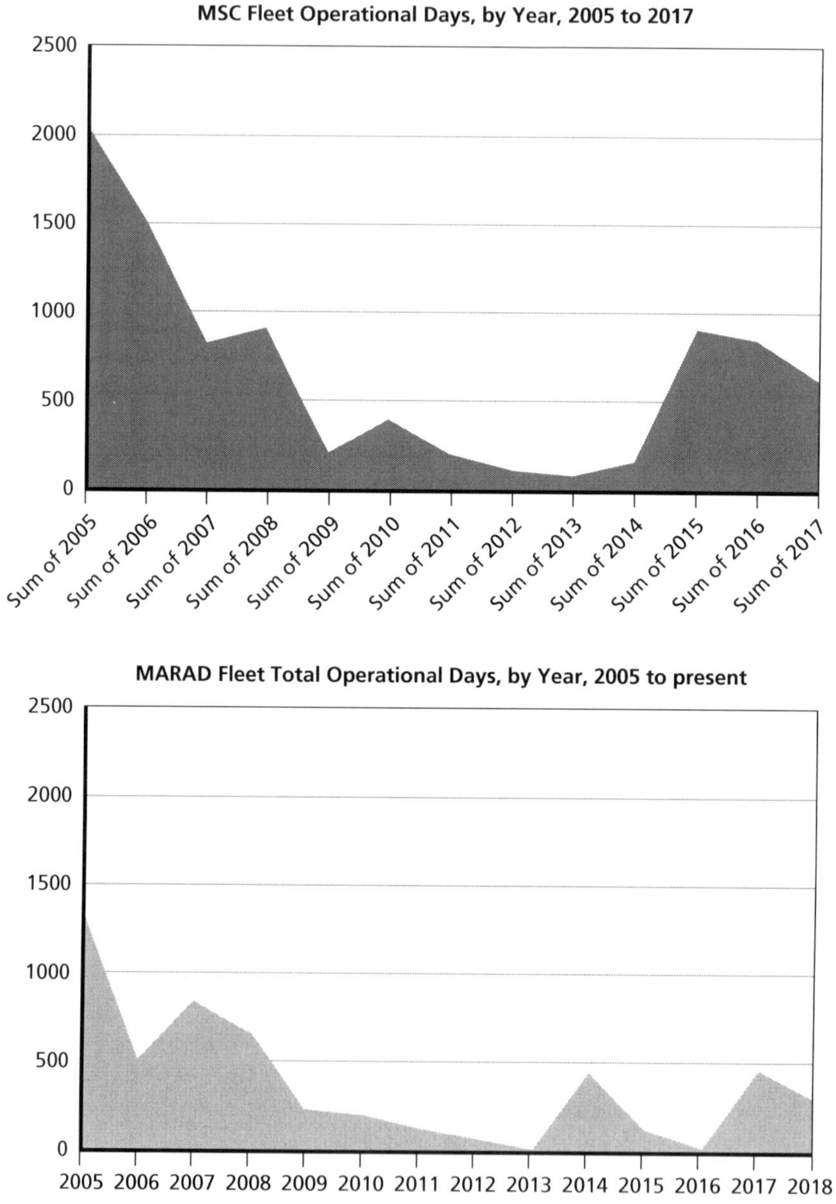

SOURCE: RAND analysis based on data provided by MSC and MARAD.

decline in use. The sealift fleet did not become smaller during this period, and there were no reported cases where a sealift ship could not be made available for a mission. Commands have continued to need these vessels in conducting operations, but not at the levels required in 2005.

During this period, the missions have varied. Strategic sealift ships have been activated for response to real-world missions ranging from disaster relief to disposal of Syrian chemical munitions. While the operations vary in length, they average about 133 days, with some being months longer. A few ships account for most of the mission activations, and some ships have not been activated in many years. Figure 2.2 breaks down the specifics of these deployments, showing that the primary uses for, in this case, MARAD's RRF were for exercises and humanitarian assistance. We did not have similar compiled data for MSC. By account only, the operational history for MSC surge sealift is more limited, consistent with a smaller number of ships, but very similar in terms of the types of operations supported.[1] There is no indication with either fleet that ships were not available to support the missions requested. In fact, the strong suggestion is that there are considerably more ships in the sealift fleet than has ever been necessary for support of limited operations.

Wartime Operational Requirements

A ship activated in peacetime is likely to be only one of a few activated among many available. There will generally be no competitors for repair and personnel resources. Assuming that some command is willing to provide the funding via TWCF, there will likely be no impediments to completing the mission, as the operational record has shown so far.

However, wartime requirements are likely to involve the activation of multiple units, not just one or two, and the resources required for this activation will be substantially more extensive. The ability to move resources from one ship to another to make a ready ship will be greatly limited. And many different units will be competing for the same resources.

Required Operational Capability / Projected Operational Environment

Ships in the sealift have an approved set of required operational capability (ROC) / projected operational environment (POE) that is intended to specify the kinds of missions these ships are required to conduct and the environment in which these missions are likely to take place. ROC/POE provides the basis for investment and readiness priority.[2] ROC/POE is not specifically a readiness requirement but a statement of needed capabilities. However, ROC/POE does have a readiness implication. If a mission is in

---

1 Interview with MSC representatives, Headquarters, Military Sealift Command, Norfolk, Va., October 26, 2018.

2 OPNAV Instruction 3501.199C, *Required Operational Capabilities (ROC) and Projected Operational Capabilities (POE) for Strategic Sealift Ships to Include the T-AKR Fast Sealift Ships (FSS), Large, Medium-Speed RO/RO (LMSR), Aviation Support Ships (T-AVB), Auxiliary Crane Ships (T-ACS), and Ready Reserve Force (RRF) Dry Cargo Ships,* U.S. Department of the Navy, Washington, D.C., 2007.

Figure 2.2
Specific Operations for Ready Reserve Force Ships

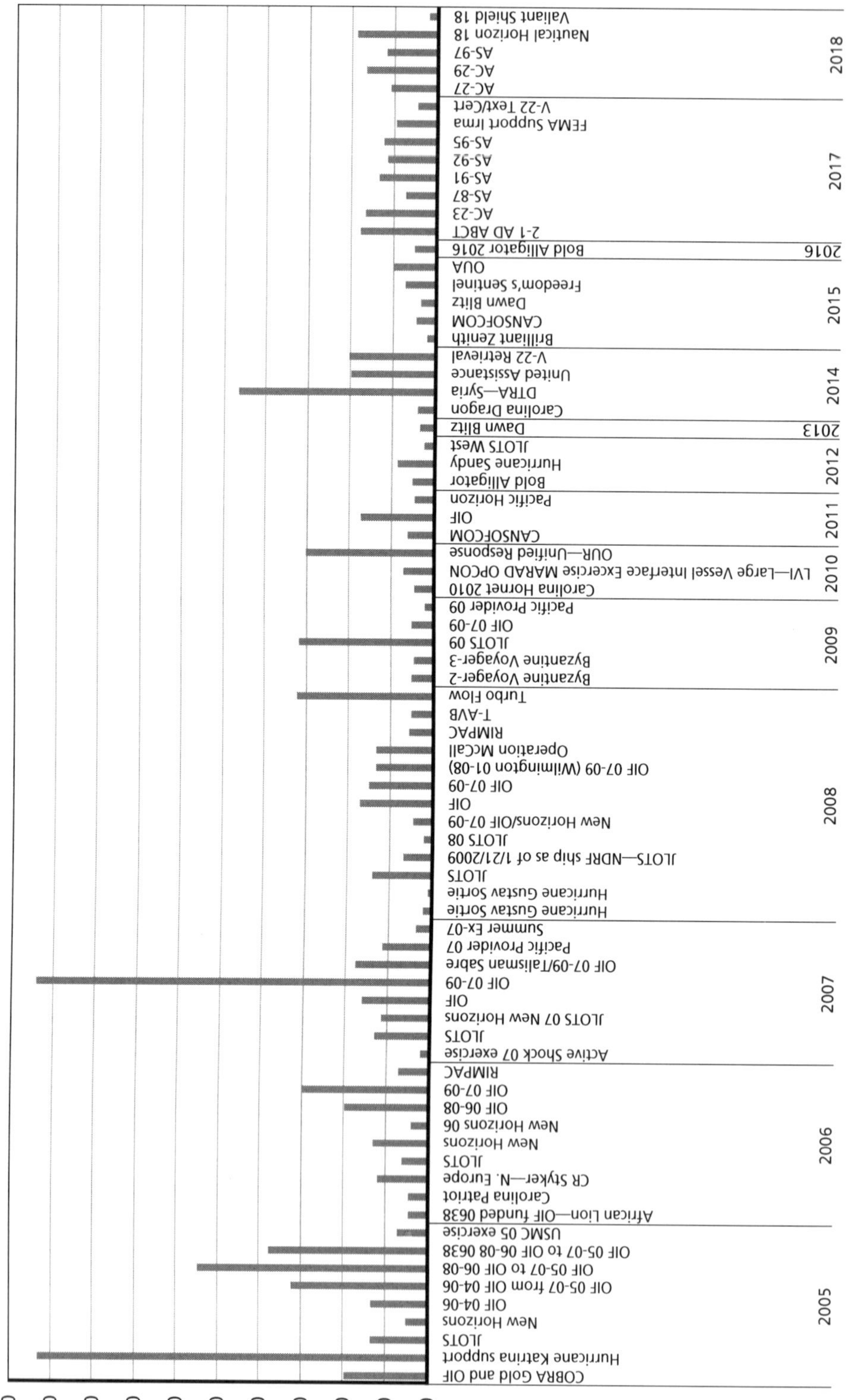

SOURCE: RAND analysis based on MARAD operational history data.

ROC/POE, the mandate is that the ship be able to perform it, and as a result, effort and resources are applied to meeting these missions.

During our ship visits, shipmasters and chief engineers mentioned that the ROC/POE for their ships did not actually match the missions assigned, with the result that effort was being expended to maintain capabilities that were not actually being used. This was particularly the case for ships that had been used for long-term prepositioning (PREPO) but had been repurposed for use in sealift. Many of these documents are not updated for current ship employment and thus impose readiness requirements that no longer apply in the ship's new employment.

For example, because PREPO ships keep equipment aboard for an extended period (approximately a year), this equipment must be stored in a climate-controlled environment to ensure the equipment is ready to function after emerging from extended storage. Maintaining these climate-control systems is an expensive and complicated task. Sealift ships embark and debark cargo within a few days or weeks and thus do not have to keep cargo holds and vehicle decks climate controlled. Thus, the climate-controlled capability is not necessary. However, as long as this capability is included in ROC/POE, it must be kept operational and ready to demonstrate to inspection teams whether or not the capability is needed for current missions.

Another capability that is similarly maintained is cargo fuel delivery. This capability is essential to PREPO ships that might debark cargo in undeveloped ports, but it is not necessary for strategic sealift. It involves, moreover, extensive maintenance and upkeep in piping systems and tanks. This capability has in fact become an expensive challenge for ships that clearly do not need the capability.

Revising ROC/POE is a long and complicated process, which is to a degree understandable given the service life of these ships and the difficulty of restoring capability once it has been abandoned. But keeping expensive requirements that are no longer needed levies a readiness bill that affects the availability of strategic sealift vessels. A more flexible system for updating ROC/POE seems essential.

Operations Plans Requirements

The RRF and surge sealift forces are sized to meet stressing operational contingencies. The RRF budget is based on the conclusions of the 2005 *Mobility Capabilities and Requirements Study 2016* and subsequent requirements review and determination by Navy and USTRANSCOM.[3]

The current funding levels are expected to support readiness and allow the ships to activate in time to deliver cargo to a given area of operations and satisfy combatant commanders' critical warfighting requirements. No specific number or pacing operation has been specified. Actual numbers and schedule for wartime requirements are

[3] U.S. Department of the Navy, Department of the Navy Fiscal Year (FY) 2018 Budget Estimates, Justification of Estimates, May 2017, National Defense Sealift Fund.

classified, so we will be discussing readiness levels in a generalized way. However, the community adopted a standard that 85 percent of the ships in the strategic sealift fleet would be available within five days of expected activation.

All plans requiring sealift are phased over several weeks, and none presume that the cargo carried on sealift needs to be moving within five days of an operations plan TPFDD being executed. Time lines vary, but in general they stretch over a period of weeks and months, not days. Indeed, there is likely not sufficient port or transportation capacity to receive all the cargo that would be moved on strategic shipping if all that were attempted within five days of operations plan TPFDD activation. As a result, the 85 percent readiness standard is neither realistic nor required.

However, there is reason to doubt that the fleet could respond to an even less demanding standard. No large-scale activation of ships has taken place since Operation Iraqi Freedom's initial phases in 2003, where 50 ships were activated over a period of several months.[4] The only mechanism currently available for assessing whether the standard could be met is the TA process, a process by which ships are selected for activation and are expected to meet the five-day window. As we will discuss in more detail, there is reason to be concerned about whether the TA process will provide an accurate picture of overall readiness.

Turbo Activations

The sealift fleet is maintained in a reduced operating service (ROS), with a requirement that the vessels must be able to get underway in five days, referred to as ROS-5. Periodically, USTRANSCOM conducts a test of these vessels' ability to activate through a TA, which is a no-notice activation. When a vessel is directed to conduct a TA, all aspects of a vessel's activation must be completed within the five-day readiness period.

We examined the history of TAs for the surge sealift fleet in a database provided by USTRANSCOM. While the number of vessels in the RRF has been relatively constant, with some vessels being phased out of the fleet and some brought in, the number of TAs has varied widely over time. The peaks and valleys of TAs have ranged from a high of 26 activations in a single year, where 26 were done at one time in 2018, to a low of 6 activations in 2010, as shown in Figure 2.3.

We next examined the frequency with which each vessel had been turbo activated to determine whether there was an equal distribution of the vessels that support a TA. The data show that while 25 vessels have been turbo activated three or more times during this time period, 28 vessels have had two or less activations in the nine-year period from 2010 to 2018 and an additional eight vessels were not turbo activated during this time. Figure 2.4 illustrates the number of TAs by vessel and whether the

---

[4] U.S. Department of the Navy, Military Sealift Command Public Affairs, "Sealift for Operation Iraqi Freedom Is No Small Thing," Washington, D.C., March 2, 2004.

Figure 2.3
Number of Turbo Activations for the Ready Reserve Force over Time

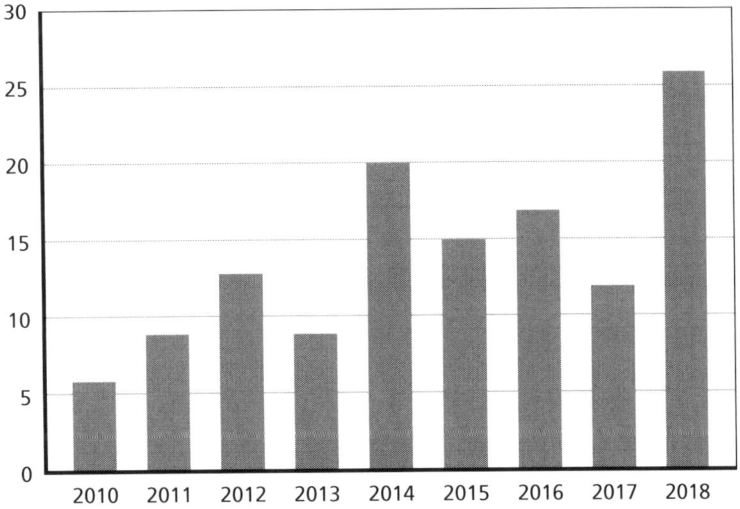

SOURCE: USTRANSCOM historical TA data.

TA was successful or unsuccessful. It is also worth noting that many RRF ships have not been activated for several years.

A more equal distribution of TAs would provide a more reliable indicator of the entire fleet's readiness to activate and support fleet requirements. These results are not meant to suggest that ships whose capabilities are primarily oriented toward wartime employment should not be kept in ready reserve or surge status. However, as we discuss in the next section, current operations do not require the force structure provided or readiness levels prescribed. Arriving at these requirements will take a review of wartime operational need.

TAs of sealift ships have been mostly successful, although there are specific problem areas within MSC's surge sealift fleet. Figure 2.5 illustrates, by year, the number of activations for MSC and MARAD. The figure further shows the success rate for each fleet, by year, and shows the general trends for both MSC and MARAD for completion of TAs. Both MARAD and MSC TA success has peaks and valleys; MARAD has generally been more consistent; MSC has recently experienced problems with specific classes, specifically Large, Medium-Speed Roll-On / Roll-Off ships that had previously been classified as PREPO vessels.

These TA results are in some respects misleading. While the TAs are supposed to reflect the readiness of ships selected on a random basis, based on the statements of crews and those performing the selections, MSC and MARAD both select ships that are most likely to successfully complete the assessment. Moreover, the TAs generally occur in isolation, and resources are routinely pooled to get ships through their

Figure 2.4
Number of Successful and Unsuccessful Turbo Activations for Sealift Fleet, 2010–2018

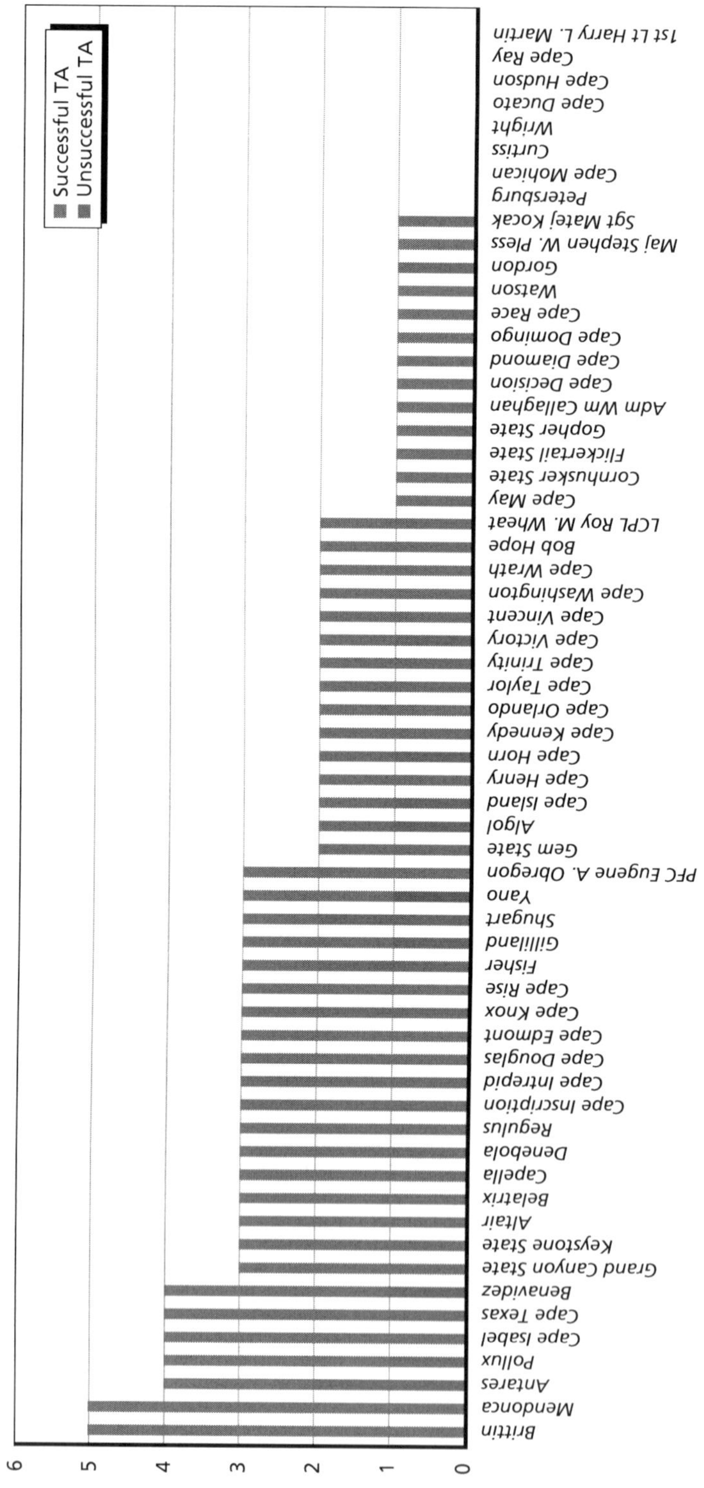

SOURCE: RAND analysis of information provided by USTRANSCOM.

Figure 2.5
Maritime Administration and Military Sealift Command Number of Turbo Activations and Success Rate, 2010–2018

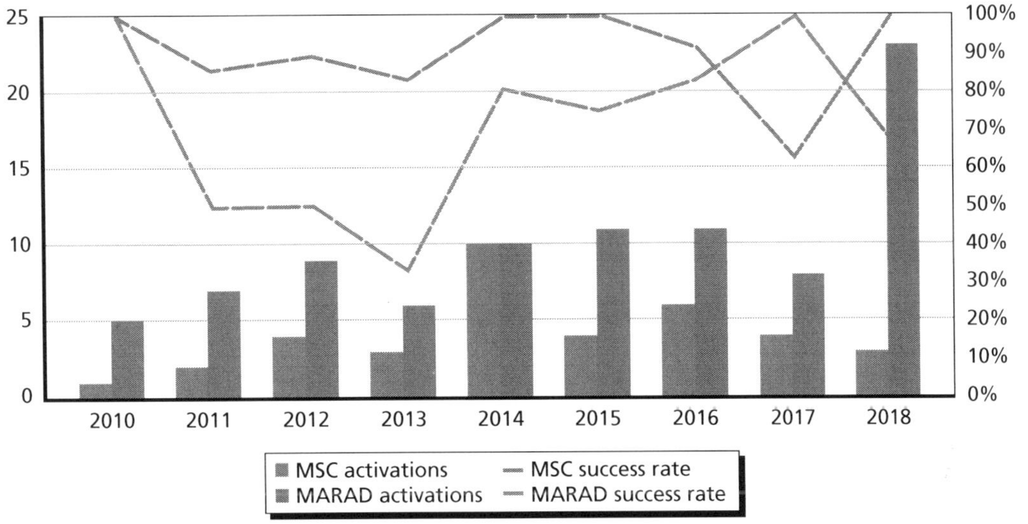

SOURCE: Data provided by USTRANSCOM.

activations,[5] even though such pooling would clearly be impractical in a large-scale activation. Taken together, all this suggests that the TA process as a means for establishing actual readiness for more than a limited number of vessels is suspect.

During our ship visits and discussions with senior RRF leaders, they conveyed that a TA is not a true test of a ship's ability to perform its mission. A better indicator would be to perform a mission. For example, an RRF vessel would be activated, complete all necessary requirements to deploy, get underway, transit to port of embarkation, and fully utilize the equipment required during an actual mission. This test would provide many benefits including a full test and run time for the vessel's engineering plant, experience for the crew, and a test of the full range of the platform's capabilities for which it is designed to operate. The trade-off is resources—it costs money and competes with other needs.

Conclusions About Wartime Readiness

There is nothing that proves that ships would not be ready within five days after a large-scale activation, but several factors suggest that this goal is unlikely, and indeed

[5] Interview with MSC personnel, October 26, 2018; Interview with MARAD personnel, Headquarters, Maritime Administration, Washington, D.C., October 11, 2018.

nothing proves that it would occur. The inability to definitively state that an activation would be successful applies even if the activation period is extended past five days and includes many ships. We will discuss reasons for doubt in greater detail in following chapters. For example, the overall shortage of merchant mariner crews is well documented, with longer-term rotations at particular risk. But an even more important unknown is the material condition of the many ships that have not recently been activated. Even among those ships that have been activated, the TAs have not been universally successful and for some ship classes notably unsuccessful. Even allowing for the generally high success rate, this rate is achieved by allowing selection based on ships likely to pass and allowing resources to be pooled. An activation of multiple ships has not been attempted, until very recently, and there is much to suggest that such an activation would be very challenging.

CHAPTER THREE

Material Readiness

The major component in reported overall readiness is material readiness. Although other readiness factors certainly affect material readiness, the major factor in determining whether a ship meets the requirement to be underway within five days is whether it has sufficient operable equipment to do so.

Sealift ships are specialized vessels, some of which have been in service for far longer than the normal service lives of merchant ships. Figure 3.1 shows the age of ships in the surge sealift and RRF fleets. None of the ships are newer than 14 years old, a few are greater than 50 years, and 23 are 45 to 49 years old. While there is no inherent reason why ships of this age cannot be maintained, there are bound to be challenges, particularly if some of the equipment on board is similarly very old. Some equipment may be several generations removed from systems in service, and, indeed, the manufacturers may be long out of business. So, based on age alone, we would expect challenges to maintain material readiness.

We examined TA data to assess potential relationships between ship age, home port location, and TA failure (or success). In Figure 3.2, the legend shows the vessel's age, name, and fleet assignment with the home port locations along the x-axis. The data indicate that some ships that were relatively younger had failed TAs, as have older ships. Ships home ported in Bremerton, Washington, and in Violet, Louisiana, each had two ships failing TAs twice. Of the four ships that failed TAs twice, two were 38 years old at the time of the failure. However, as we've noted in the earlier discussion of TAs, many vessels never receive TAs, so this evidence does not by itself validate a conclusion of an association between age and material condition. There is, however, an abundance of anecdotal evidence that older ships present maintenance challenges that would affect ship readiness. Recent discussions concerning recapitalization of the fleet also support this assessment.[1] There was also anecdotal evidence that it is difficult to find qualified mariners to man ships in Violet, Louisiana. In sum, there are numerous factors that may affect the readiness of vessels and others—including age, home port, location, and manpower—that affect the readiness of the fleet.

---

[1] U.S. Government Accountability Office, *Navy Readiness: Actions Needed to Maintain Viable Surge Sealift and Combat Logistics Fleets*, Washington, D.C., GAO-17-503, 2017.

Figure 3.1
Age of Vessels in Surge Sealift and Ready Reserve Force

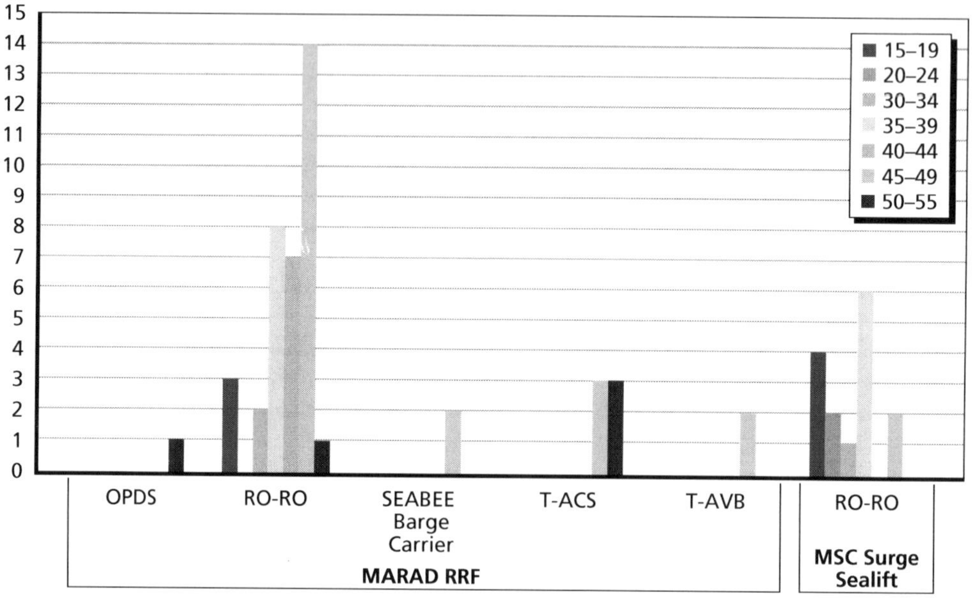

SOURCE: MSC and MARAD data calls.
NOTES: OPDS = Offshore Petroleum Distribution System; RO-RO = Roll-On/Roll-Off Ships; T-ACS = Auxiliary; T-AVB = Aviation Support Ships.

Figure 3.2
Ages of Ships Failing Turbo Activations, 2010–2015

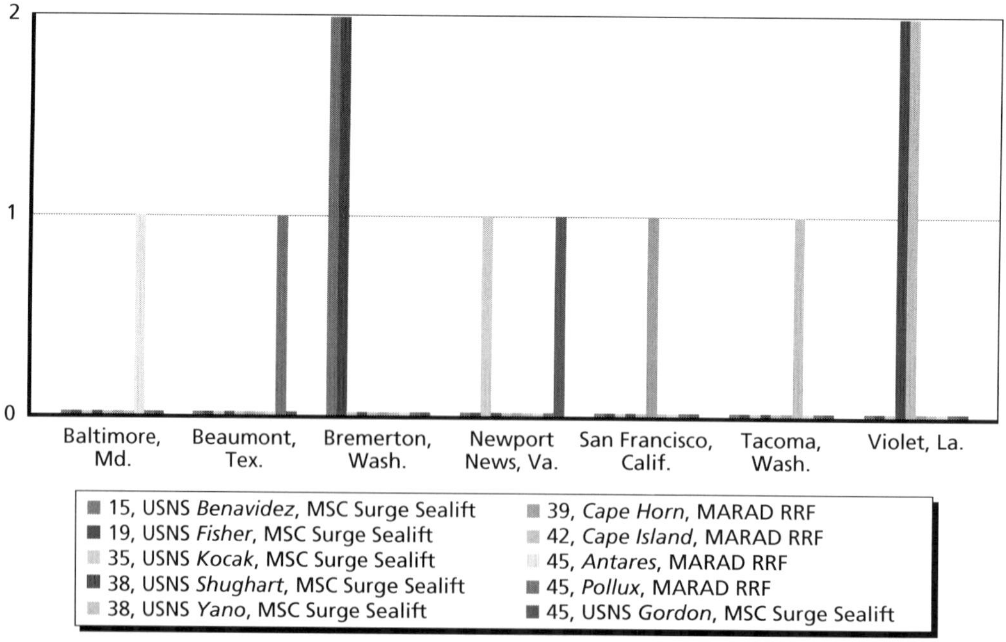

SOURCE: USTRANSCOM.

Among the crews interviewed for this study, there was considerable frustration at the lengthy period and documentation required to upgrade to new equipment on old ships. These individuals described the TRANSALT process (an approved alteration that replaces an originally installed system with a newer one) as cumbersome to the point that old equipment with spare parts long out of production remained on ships, sometimes imposing long repair delays.[2] We will discuss different management approaches in a later chapter, but this concern further illustrates the challenges faced in keeping old ships with old systems operational.

Organizational-Level Maintenance

Organizational-level maintenance refers to work that crews conduct on the ships where they are assigned. This work includes planned maintenance, material preservation, and simple repair and is conducted by ROS crews assigned to each ship during inactivated status and by larger full operational status (FOS) crews when the ships are activated. We will discuss in the next chapter the details of ROS and FOS crews. For this chapter, we will focus on the kind of maintenance these crews are responsible for conducting throughout the ship's inactivated periods.

Organizational-Level Maintenance During Inactivation Periods

During inactivated periods, ship managers hire crews who work to keep the ship ready for activation within the five-day required window. These crews are sized specifically for performance of the kind of maintenance viewed as feasible for periods in which no equipment is operating. Table 3.1 is drawn from the maintenance protocol document used by MARAD and shows the kinds of work expected to be conducted at various periodicities. Crews perform these maintenance actions, as well as required repairs that are within their capacity and expertise to do so. Performance of these actions is then reported by contracted ship managers to platform managers of the various ship types. MARAD[3] and MSC[4] have very similar procedures.

Between these two organizations, there is no report of missed organizational level maintenance or inability of crews to meet the demands. However, there are only two ways under current practices to verify maintenance performance: through activations (either TAs or real-world operational activations) or through U.S. Coast Guard (USCG) / American Bureau of Shipping (ABS) certification currency. We know that real-world activations occur on a narrow range of ships and enjoy the full attention of the activating organization when they occur. If failures of organizational mainte-

---

[2] Interview with MSC personnel, October 26, 2018; Interview with MARAD personnel, October 11, 2018.

[3] Management Systems Consultants, *MARAD NS5 Maintenance Protocol*, Sugar Land, Tex., May 2016.

[4] M. H. Buzby, *The U.S. Navy's Military Sealift Command Standard Operating Manual*, U.S. Department of the Navy, Washington, D.C., COMSCINST 3121.9C, 2012b.

Table 3.1
Organizational Maintenance Periodicities

Maintenance Category/R-Status	ROS-4/5/10	RRF-10	RRF-20	RRF-30
Outported	Yes[c]	No[d]	No	No
ROS crew	Yes	No[e]	No	No
Maintenance activation with dock trial	Annual[a]	None	None	None
Maintenance activation with sea trial	Twice in 5 years[b]	Twice in 5 years[b]	Once in 5 years	Once in 5 years
Planned maintenance cycle	Continuous	6 months	1 year	1 year
Vibration analysis	Twice in 5 years[b]	Twice in 5 years[b]	Once in 5 years	Once in 5 years
Main diesel engine analysis	Twice in 5 years[b]	Twice in 5 years[b]	Once in 5 years	Once in 5 years
Lube oil analysis	6 months	6 months	At activation[f]	At activation[f]
Diesel engine crankshaft deflection	Annual[a]	Twice in 5 years[b]	Once in 5 years	Once in 5 years
Cylinder drain lube oil analysis	Twice in 5 years[b]	Twice in 5 years[b]	Once in 5 years	Once in 5 years
Infrared photographic thermography	Twice in 5 years[b]	Twice in 5 years[b]	Once in 5 years	Once in 5 years
Insulation resistance (low voltage) and Megger test (high voltage)	Annual	Annual	Once in 5 years	Once in 5 years
Diesel engine scavenge port inspection (2 stroke propulsion engines only)	Twice in 5 years[b]	Twice in 5 years[b]	Once in 5 years	Once in 5 years

SOURCE: MARAD NS5 Protocol, May 23, 2016.

[a] Dock trials to be scheduled in non–sea trial fiscal years. These dock trials can be used to support regulatory requirements.

[b] Sea trials twice in 5 years (one post drydock and one intermediate).

[c] Generally outported, however, some may be located at NDRF sites.

[d] Generally located at NDRF sites; however, some may be sited at outport locations.

[e] Generally without crew; however, some vessels may have a small retention crew.

[f] MARAD, at its own discretion, may elect to do some equipment (e.g., stern tube) more frequently.

nance occurred, those would be corrected in the course of getting the ship ready to deploy, and the fact of missed maintenance up to that point might not receive much attention. We also know that TAs are similarly biased in selection and scope. Because resources are pooled to ensure successful completion of a TA, missed maintenance up to that point would be corrected and the fact of missed maintenance might not receive much attention.

However, USCG/ABS certifications show a different pattern. Under agreement with ABS, USCG delegates responsibility for certifications to ABS. ABS conducts these assessments at intervals, and completion of these assessments is a condition for operations. Failure to complete these certifications can be a matter of teams not being available or substantial and unexpected casualties. However, missing certifications can indicate that some maintenance is not being performed, inspections are not being managed appropriately, or the database is in error. The currency of inspections and certifications is an indicator of material readiness. USCG maintains the Port State Information Exchange (PSIX) database,[5] updated weekly, which provides the status and currency of vessel certifications and inspections. MSC and MARAD surge sealift vessels are reported in the database. Table 3.2 contains the titles of the inspections/certifications, their purpose and periodicity, and the reference requiring the inspection.

We examined the entire fleet's currency for the required inspections. Tables 3.3 and 3.4 illustrate the current certification status for MARAD and MSC vessels, respectively, on June 28, 2018, when the USCG's PSIX database was accessed. Across the top row is the list of the required inspections or certifications. The left column is the vessel name. Within the table the red coloring indicates a lapsed certification, green represents current certification, and yellow indicates the certification status was not in the database.

As the figure indicates, many certifications are not current. Some ships have several certifications that have expired; the most numerous certifications that are expired are ISM Documentation and ISM Safety Management. Only a few ships are completely current with their certifications; most have several certifications that are not current, and many ships in each fleet were all missing required certifications. The inspections status recorded in these figures reflect a snapshot in time and likely does not represent the currency of inspections/certifications today.

The reasons for expired certifications may be varied but calls into question the readiness of the fleet to support full-scale activation. Expired certifications are not reported in readiness reports, as are casualty reports. However, they may be indicative of the readiness of a vessel to sail. Ships with a large number of expired certifications may be at risk of safely employing the vessel. Even a few outstanding inspections may affect the ability of a vessel to rapidly activate, depending on what is needed to bring the vessel current. The time, personnel, financial resources, and logistics needed to bring inspection requirements current may be large. USTRANSCOM, MARAD,

---

[5] The PSIX system contains vessel-specific information derived from USCG's Marine Information Safety and Law Enforcement System (MISLE). The information contained in PSIX represents a weekly snapshot of Freedom of Information Act data on U.S. flag vessels, foreign vessels operating in U.S. waters, and USCG contacts with those vessels. Information on unclosed cases or cases pending further action is considered privileged information and is precluded from the PSIX system. U.S. Coast Guard Maritime Information Exchange, "Port State Information Exchange," webpage, January 28, 2019.

Table 3.2
Vessel Inspection and Certification Requirements

Inspection/Certification	Definition/Purpose	Periodicity	Reference
USCG Certification of Inspection (COI)	Describes the vessel, the route it may travel, minimum manning requirements, survival and rescue craft, general safety equipment	5 years	Code of Federal Regulations (CFR) 46, Part 115, Subpart A.
USCG Certificate of Documentation (COD)	National registration of vessels which demonstrates ownership of vessel	Annual	Vessel Documentation Online, homepage, 2018.
Classification Document International Load Line	Certifies the correctness of load line marks and that the vessel is in compliance with applicable requirements; also describes the load line marks, conditions, restrictions, and exemptions	5 years	CFR 46, Part 42, Subpart 42.07.
Oil Pollution Prevention Certificate	Complete examination of structure, equipment, systems, fittings, arrangements, and material	5 years	CFR Subchapter O, Part 155/157 s; CFR 33, Part 151, Subpart A.
Safety of Life at Sea (SOLAS) Cargo Ship Safety Construction	Examination of the condition of the structure, machinery and equipment as defined in regulation was satisfactory and the ship complied with the relevant requirements	Renewal survey required maximum of 5 years	Safety of Life at Sea (SOLAS) 1/10, II/1, II/2; SOLAS 1974, regulation I/12; 1988 SOLAS Protocol, regulation I/12.
SOLAS Ship Safety Equipment Certification	Check that there are adequate amounts of usable life-saving appliances (boats, jackets, etc.)	A renewal survey; maximum of 5 years	SOLAS 1974, regulation I/12; 1988 SOLAS Protocol, regulation I/12.
Ship Safety Radio Certification	Survey of radio equipment, installation, operation on ship as well as the radios in survival craft	Not to exceed 5 years	SOLAS 1974, regulation I/12, as amended by the GMDSS amendments; 1988 SOLAS Protocol, regulation I/12.
ISM DOC	Certifies that the safety management system has been audited and complies with requirements of ISM code	5 years	SOLAS 1974, Chapter 9
ISM Safety Management	Ship-by-ship safety management system audit	5 years	SOLAS 1974, Chapter 9

and MSC need to identify challenges related with meeting certification requirements and plan accordingly. Expired certifications are not solely due to inability to perform organizational-level maintenance. Scheduling alone can create shortfalls, as can unexpected casualties that are not due to maintenance lapses. However, expired certification does suggest issues with the management of ROS-5 ships that are not being captured in TA performance or elsewhere.

Table 3.3
U.S. Coast Guard Certification Status for Ready Reserve Force Vessels

	USCG COI	USCG COD	Classification Document	International Load Line	Oil Pollution Prevention Certificate	SOLAS Cargo Ship Safety Construction	SOLAS Ship Safety Equipment Certification	Ship Safety Radio Certification	ISM DOC	ISM Safety Management
Cape Domingo	O	O	O	O	O	O	O	O	X	X
Cape Diamond	O	O	E	O	E	E	E	E	E	E
Cape Douglas	O	O	E	O	E	E	E	E	E	E
Cape Ducato	O	O	E	E	E	E	E	E	X	X
Cape Edmont	O	O	E	E	E	X	X	E	X	X
Cape Race	O	O	O	O	O	O	O	O	O	O
Cape Ray	X	O	O	O	O	E	O	O	E	E
Cape Rise	O	O	O	O	O	O	O	O	O	O
Cape Washington	O	O	O	O	O	E	O	O	X	X
Cape Wrath	O	O	O	O	E	O	E	E	E	E
Cape Kennedy	E	O	O	O	O	O	O	O	E	E
Cape Knox	E	O	O	O	O	O	O	O	E	E
Cape Taylor	O	O	O	O	O	O	O	O	O	O
Cape Texas	O	O	O	O	O	O	O	O	E	E
Cape Trinity	O	O	O	O	O	O	O	O	O	E
Cape Victory	O	O	O	O	O	O	O	O	O	O
Cape Vincent	O	O	O	O	O	O	O	O	O	O

Table 3.3—Continued

	USCG COI	USCG COD	Classification Document	International Load Line	Oil Pollution Prevention Certificate	SOLAS Cargo Ship Safety Construction	SOLAS Ship Safety Equipment Certification	Ship Safety Radio Certification	ISM DOC	ISM Safety Management
Adm W Callaghan	O	O	X	O	E	E	E	X	E	E
Cape Henry	O	O	O	O	O	O	O	O	O	O
Cape Horn	O	O	O	O	O	O	O	O	X	E
Cape Hudson	O	O	E	E	E	E	E	E	E	E
Cape Inscription	O	O	O	E	O	E	O	O	E	E
Cape Intrepid	O	O	E	E	E	E	E	E	E	E
Cape Isabel	O	O	O	O	O	O	O	O	O	O
Cape Island	O	O	O	E	O	O	O	O	E	E
Cape Orlando	O	O	O	O	O	O	O	O	E	E
Cape May	O	O	O	O	O	O	O	O	E	E
Cape Mohican	O	O	E	O	O	O	O	E	E	E
SS *Antares*	O	O	E	E	E	E	E	X	E	E
SS *Denebola*	O	O	O	O	O	O	O	O	E	E
SS *Bellatrix*	O	O	O	O	O	O	O	O	O	O
SS *Pollux*	O	O	O	O	O	O	O	O	O	O
SS *Regulus*	O	O	O	O	O	O	O	O	E	E
SS *Algol*	O	O	O	O	O	O	O	O	E	E
SS *Capella*	O	O	O	O	O	O	O	O	E	E

Table 3.3—Continued

	USCG COI	USCG COD	Classification Document	International Load Line	Oil Pollution Prevention Certificate	SOLAS Cargo Ship Safety Construction	SOLAS Ship Safety Equipment Certification	Ship Safety Radio Certification	ISM DOC	ISM Safety Management
SS *Cornhusker State*	O	O	E	E	E	E	E	E	E	E
SS *Flickertail State*	O	O	O	O	O	O	O	O	E	O
SS *Gopher State*	O	O	O	O	O	O	O	O	O	O
SS *Gem State*	O	O	E	E	E	E	X	E	X	X
SS *Grand Canyon State*	O	O	O	O	O	O	O	O	O	O
SS *Keystone State*	O	O	O	X	E	E	E	X	X	X
SS *Wright*	O	O	O	O	O	O	O	O	E	E
SS *Curtiss*	O	O	O	O	O	O	O	O	O	O
SS *Petersburg*	O	O	O	O	E	X	E	E	O	O

SOURCE: USCG Port State Information Exchange (PSIX) database.

NOTE: Certification status as of June 28, 2018. E (red) means the certificate is expired in PSIX database; X (yellow) means that the certificate was not located in the database; O (green) signifies that the certificate is up to date.

Table 3.4
Certification Status for Military Sealift Command Surge Sealift

	USCG COI	USCG COD	Classification Document	International Load Line	Oil Pollution Prevention Certificate	SOLAS Cargo Ship Safety Construction	SOLAS Ship Safety Equipment Certification	Ship Safety Radio Certification	ISM DOC	ISM Safety Management
USNS *Gilliland*	E	X	E	E	E	E	E	E	E	E
USNS *Gordon*	O	X	E	E	O	O	O	O	E	E
USNS *Benavidez*	O	X	O	O	O	O	O	O	E	E
USNS *Mendonca*	O	X	O	O	O	O	E	E	E	E
USNS *Shughart*	O	X	E	E	E	O	E	E	E	E
USNS *Watson*	O	X	O	O	E	E	O	O	O	O
USNS *Yano*	O	X	O	O	O	O	E	O	O	O
USNS *Bob Hope*	O	X	O	O	O	O	O	O	O	E
USNS *Brittin*	O	X	O	O	O	E	O	O	E	E
USNS *Fisher*	O	X	O	O	O	O	O	O	O	O
USNS *Sgt Matej Kocak*	O	X	O	O	O	O	O	O	E	O
USNS *LCPL Roy M. Wheat*	O	X	O	O	X	X	X	E	O	E
USNS *1st Lt Harry L. Martin*	O	X	O	O	O	O	O	O	E	E
USNS *PFC Eugene A. Obregon*	O	X	O	O	O	O	O	O	O	O
USNS *MAJ Stephen W. Pless*	O	X	O	O	O	O	O	O	E	X

SOURCE: USCG Port State Information Exchange database.

NOTE: Certification status as of June 28, 2018. E (red) means the certificate is expired in PSIX Database; X (yellow) means that the certificate was not located in the database; O (green) signifies that the certificate is up to date.

Depot-Level Maintenance

Although ROS crews achieve a certain amount of organizational maintenance, most maintenance and repair are performed by industrial facilities (depots). This work is done in accordance with plans developed by MARAD and MSC and is driven both by regulatory need and in response to conditions identified by crews. Depot maintenance is the major component of readiness cost and the one most requiring correct identification of requirements and efficient management execution. Failing to correctly identify and resource maintenance needs and then failing to completely and correctly execute the availabilities can result in short-term lack of readiness and longer-term diminishment of service life.

There are differences in the mechanisms by which MARAD and MSC receive funding for depot maintenance. However, both use comparable contracting mechanisms with ship repair providers. Both are drawing from the same pool of ship repair contractors, which for a variety of reasons is limited in the United States.[6]

Ready Reserve Force Maintenance

MARAD's depot maintenance expenditures are mission funded; that is, they are provided as a general appropriation to MARAD without earmarking for vessels. This mechanism gives MARAD flexibility for prioritizing the maintenance requirements of vessels within a given period. If one vessel has more maintenance needs or more pressing operations than some others, MARAD can adjust resources.

However, this flexibility does not translate into more overall resources. MARAD's overall maintenance and repair budget for sealift was effectively flat from 2013 through 2016, with an increase in 2017, as shown in Figure 3.3. Since these years include periods in which sequestration caps were present, the expenditure levels likely reflect fiscal constraint rather than the full maintenance requirements of the fleet. In fact, if we consider only the overall levels of funding provided relative to the reported readiness of the fleet, all that we can infer is that the maintenance provided was sufficient to approach the 85 percent readiness standard. If the reporting is accurate, and the TA results can be trusted as a representative sample of the overall readiness, the immediate conclusion is that MARAD is not experiencing funding shortfalls in providing depot maintenance, and the practice of prioritizing within a general appropriation is effective.

In fact, what may be happening is that the practice of shifting priorities might have been all too effective in portraying ships as ready, possibly at the expense of correctly identifying increased overall requirements. None of the ships are getting newer, and none of the maintenance requirements are going down. However, the ships that

---

6 There are strict provisions regarding the use of overseas maintenance providers, and neither organization is regularly able to access capabilities beyond those available in U.S. home ports. U.S. Code, Title 10, Section 7310, Overhaul, Repair, etc., of Vessels in Foreign Shipyard: Restrictions, December 12, 2017.

Figure 3.3
Overall Maritime Administration Maintenance and Repair Budgets

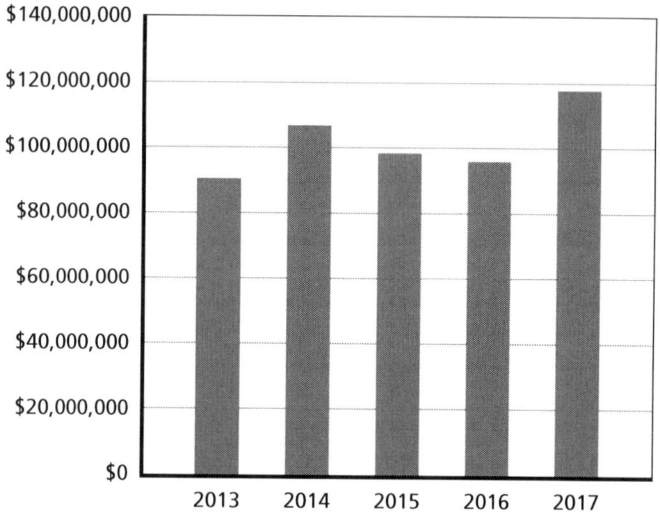

SOURCE: MARAD.

for some reason are experiencing issues get immediate attention, while those that are not being evaluated in TAs or on the cusp of operations will wait until they become a priority. This probably has a satisfactory outcome up to the point that several requirements become imminent all at once.

To support the conclusion that resources are being moved to meet immediate needs, MARAD's records (Figure 3.4) show a steep increase in known high-priority maintenance work that must be completed prior to a ship being allowed to operate but that was not originally planned to be funded in 2018 and 2019. This means that more work was identified for immediate correction than was originally budgeted. The assessment of subject matter experts is that the additional requirements might have been due to the USCG/ABS inauguration of an enhanced inspection program. These enhanced inspections were mandated by Congress in response to the SS *El Faro* sinking in 2015 in which material shortfalls played an important role.[7]

The enhanced inspection program applies to all U.S. flag merchants, including those owned by the U.S. government, and has the effect of both identifying more discrepancies and forcing action on them. Those ships receiving enhanced inspections get immediate attention. However, since the overall level of funding is not increasing, those ships not immediately in the certification window are likely not receiving more

[7] World Maritime News Staff, "US Congress Passes El Faro Maritime Safety Act," *World Maritime News*, September 28, 2018.

Figure 3.4
Unfunded High-Priority Requirements for Maritime Administration Vessels

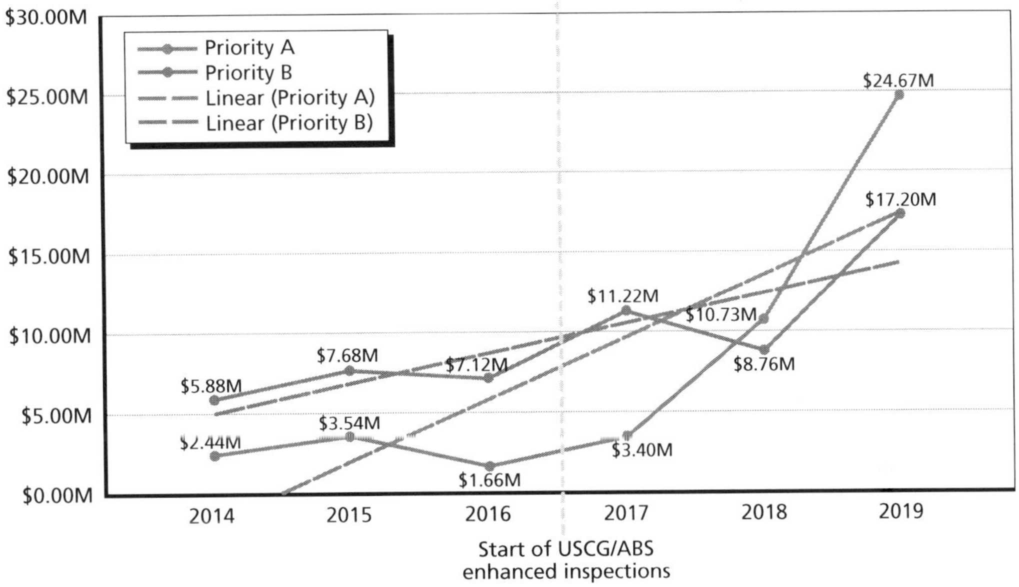

SOURCE: MARAD.

routine care. This does not necessarily result in an immediate impact on reported readiness. However, over time, it likely does mean a longer-term readiness impact as conditions that might earlier have been addressed as routine become emergent. Of note, MARAD does keep track of deferred but not emergent maintenance. Figure 3.5 shows that deferred but not emergent maintenance has also increased, although it is not as steep as the curve for emergent maintenance.

We begin with a premise that MARAD's overall readiness is sufficient to meet the requirement and that, given this premise, there is little reason to conclude funding levels are insufficient. However, it is important to bear in mind that, as explained here, this seeming success may be suppressing demand.

Ready Reserve Force Repair Contractor Performance

We received little direct documentation concerning RRF contractor performance, other than the output that RRF vessels generally approach the 85 percent readiness standard, without notable cases of RRF vessels being out of readiness for greater than the expected period. Vessels are generally in and out of their availabilities as scheduled. However, as a caveat, we do not have detailed availability completion reports, so we are not certain of what effort MARAD had to exert to reach this status. Stating that the fleet meets the readiness standard also relies on our confidence in the reporting. If there is reason to doubt that the standard is being met, there may be reason to doubt that contractor performance is as effective as it appears.

Figure 3.5
Ready Reserve Force Deferred but Not Emergent Maintenance

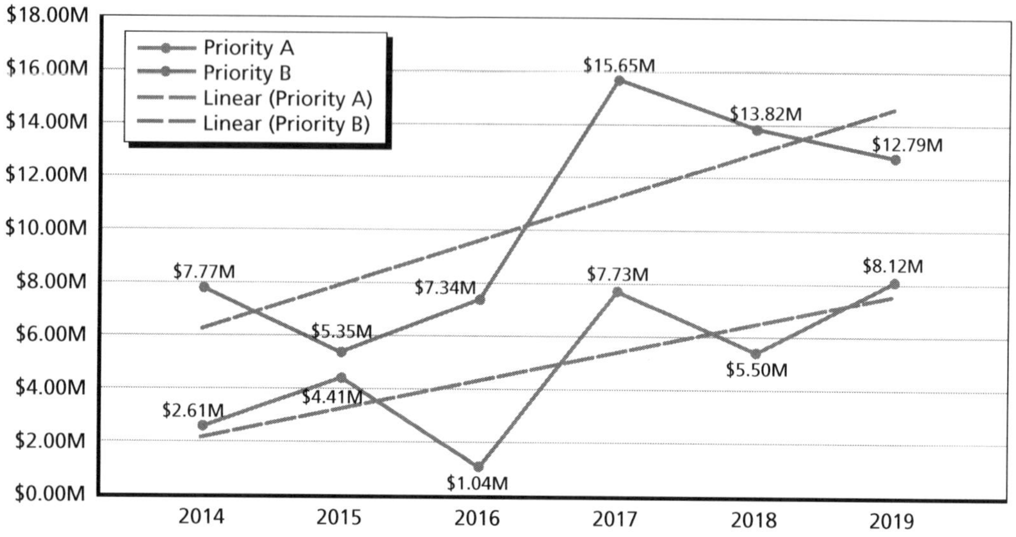

SOURCE: MARAD.

Surge Sealift Maintenance

MSC maintenance and repair is funded through a working capital fund construct, which means that rather than receiving a fixed budget for work, it is funded on a per-availability basis for work to be performed on an expected rate. While both MARAD and MSC work through contracting mechanisms with providers, MSC's maintenance and repair funding is effectively done availability by availability.

MSC's maintenance and repair budget, which includes more than sealift, has varied measurably across time, going through a period of declining resources followed by a generally upward trend from 2013 to 2017, as Figure 3.6 depicts. Of note, however, is that the executed maintenance and repair budgets for 2013–2017 showed substantial growth from budgeting to execution (Figure 3.7).

Indeed, focusing only on MSC surge sealift, maintenance availabilities in 2016 and 2017 experienced an average growth rate of 78 percent (Figure 3.8). When USNS *Gilliland*'s availability is added, the average growth is 180 percent. Since we are looking at a limited number of availabilities (four) in a limited time span, this is not necessarily a trend. However, it does indicate that the planned activity in a given year was measurably less than what was needed and that the only way to address this was through funding and completing the growth work rather than by deferring or reprioritizing the maintenance.

Figure 3.6
Military Sealift Command Overall Maintenance and Repair Budget

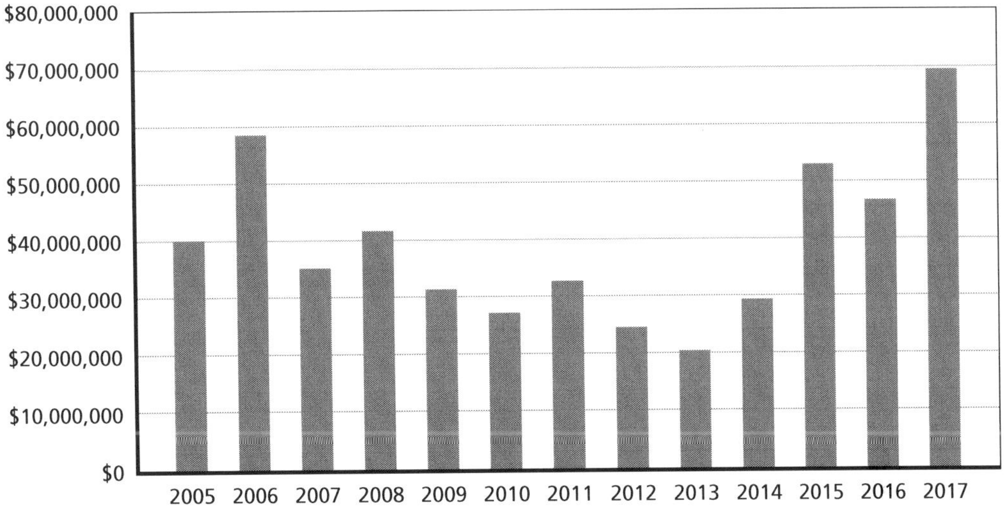

SOURCE: Budget exhibits.

Figure 3.7
Military Sealift Command Planned Versus Executed Maintenance and Repair Budget Levels

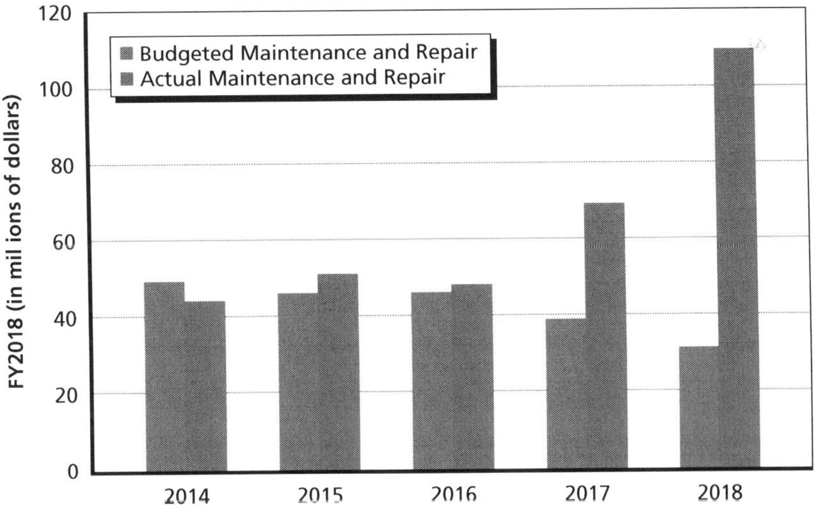

SOURCE: MSC budget data.

Figure 3.8
Growth in Military Sealift Command Surge Sealift Operations

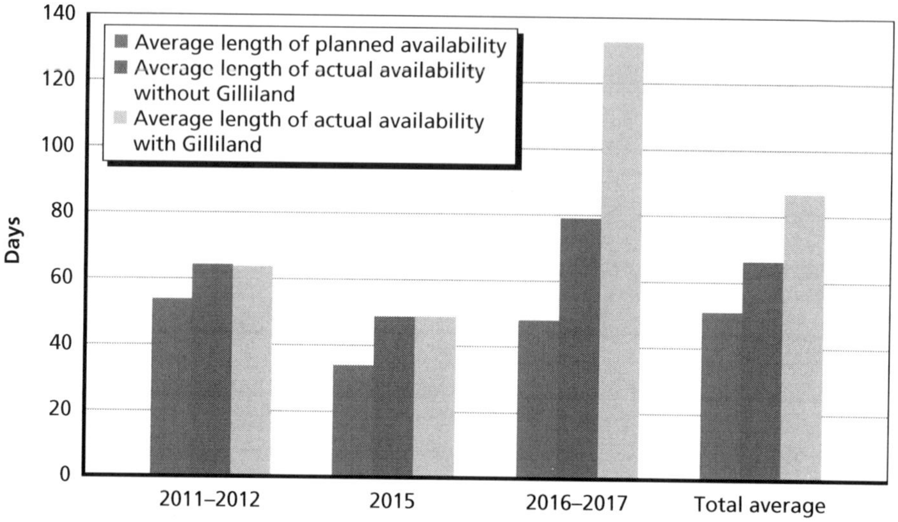

SOURCE: MSC.

Surge Sealift Repair Contractor Performance

MSC surge sealift was facing the same enhanced USCG/ABS inspection regime as MARAD, so this may account for some of the growth in 2016 and 2017 availabilities. However, this very high rate of growth does bring into question both the process for identifying work and the performance of the contractors executing the availabilities. We received 32 departure reports for availabilities completed on surge sealift vessels dating to 2006, which represents just less than half of the total availabilities performed on surge sealift during this period.[8] The reports show that every recorded availability in this period experienced growth of 10 percent in cost and schedule. The reports generally do not include any detail on reasons for cost and schedule escalation.

During discussions with subject matter experts, some of their comments bear on repair contractor performance and apply to a variety of contractors. The ones reported below are the most critical, and we have omitted the names of the contractors, as we do not have independent verification of the comments. However, these do suggest places where MSC itself identified problems both with process and execution:

- *Work not identified*: "The severity of the hull pitting and the extent of rudder repairs required contract extensions and an enormous amount of coordination between the shipyard, the shipyard's contractors, and the ship's force in order to complete."

[8] We received no departure reports from MARAD, so this is not intended as comparison.

- *Poor quality assurance*: "The change order pricing was a constant battle. The pricing was normally 3 to 4 times the value of the repair."
- *Workforce instability*: "The shipyard laid off its own work force, and yard relied almost entirely on sub-contractors."
- *The availabilities showing the largest growth do not have detailed reasons provided*: "However, the list of work completed does indicate that growth due to corrosion might have been a major factor, i.e., work not identified."

The reports do include the 2017–2018 USNS *Gilliland* availability with Bayonne Drydock and Repair as an outlier. Its availability went to 401 days (into 2018) and experienced financial growth from $28,675,236 originally awarded to $83,062,000 actually executed, a growth percentage of 572.45 percent.[9] Beyond that, the departure report does not include much detail on why the availability went for such an extended period at such an extended cost.

We did not have a comparable set of departure reports for the MARAD administered availabilities, and we are faced with two different processes and policies for dealing with these events. We consequently cannot say whether the critical comments applied by MSC to some providers are generally applicable.

Depot Maintenance Summary

While we can say that RRF material readiness generally seems better than surge sealift material readiness, this is based largely on reporting, which in each case is against a set of different criteria. Some of this may also be due to MSC being responsible for many ships that previously served in the PREPO force, ships that have been particularly challenged to keep certifications and pass TAs.[10]

However, it is clear that for both the RRF and surge sealift, there is evidence that conditions exist that were either identified late in the availability planning process or were identified but deferred. MARAD appears to be meeting this challenge by managing the overall fleet, sometimes moving resources to higher-priority availabilities, while MSC is obliged to start and finish individual availabilities without the ability to reschedule or move resources. Both approaches have advantages and liabilities. However, the important point for both is that work necessary for sustainment, and not just for immediate repair, exists and may create a bow wave of demand, either across the force or in individual availabilities.

---

9 MSC availability departure report.

10 The research team made site visits to several ships, administered by both MSC and MARAD, and found that the "converted LMSRs," which had previously been PREPO ships, were in particularly poor material shape, an impression that was confirmed by the masters and ROS crews of these ships.

Supply Sustainment

The last piece of material readiness we will consider is supply support, which affects organizational maintenance, depot maintenance, and ability to sustain underway operations. Supply support is less focused on the provision of consumables than on the provision of repair parts for installed equipment. In considering this, it is important to remember that the sealift fleet is old relative to transoceanic merchant shipping and as a result has equipment that might not have been produced for years or decades. We did not have access to a complete record of parts ordered for RRF or surge sealift in availability or operations, so our conclusions are based on information provided by ship personnel. However, from information received in these interactions, the kinds of casualties identified as taking a long period to correct are associated with replacing equipment for which parts are scarce.

A variety of issues might cause parts to be difficult to obtain. The original equipment manufacturer might be long out of business. The equipment is old enough that suitable substitutes for the original parts are not available. The parts might have to be specially manufactured. The policy requirement to favor the original equipment manufacturer in providing service further complicates the procurement and repair process. While this practice is based on a desire to preserve industrial base, it provides little incentive for efficient production of scarce parts. The original equipment manufacturer is effectively assured of business at any price it chooses to demand and at a pace it chooses to deliver. This policy merits review.

Obsolescent Equipment and Alteration Management

Maintaining obsolescent equipment is expensive, and the cost of repair eventually exceeds the cost of replacement. This is particularly true when there are suitable substitute systems available, as frequently is the case when dealing with such equipment as air compressors or air-conditioning units or even main propulsion systems. These systems might have gone through several evolutions since a ship originally went into commission. Trying to keep the old systems operational will be expensive and might be futile.

While cost-benefit analysis might dictate outright replacement of an originally installed system with a newer one, this requires an approved alteration, or TRANSALT. The TRANSALT process is intended to preserve configuration control, and there is indeed good reason not to allow unconstrained installation of different equipment on the same ship class.[11] Multiple configurations can become difficult and expensive to sustain. However, our discussions with subject matter experts pointed to this process as cumbersome to the point that approval for alterations can take years during which time another generation of equipment might already have been developed and fielded.[12] This is another area where improved management practice might have immediately beneficial effects.

---

[11] TRANSALT process guidance.

[12] Interview with MSC personnel, October 26, 2018.

CHAPTER FOUR

Personnel Readiness

Both surge sealift and RRF vessels have crews that are there during periods of ROS.[1] These crews are augmented by larger crews when activated to FOS. A key component to surge readiness is the reliance on ability of the ROS crew to maintain and sustain the vessel. In this chapter, we address the responsibilities of ROS crews, crew training requirements and challenges, differences between ROS and FOS crews, management of ROS crews, and the supply of and demand for mariners.

What Reduced Operating Status Crews Do

ROS crews perform maintenance and sustainment work to support the activation of the vessels. Typically, ROS crews work eight hours per day, five days a week (excluding federal holidays). Normal working hours are from 0800 to 1700 local, including time allotted for breaks and lunch.

Ship's maintenance managers are responsible for hiring ROS crewmembers. The contract with the ship's maintenance manager (or contractor) delineates general and specific responsibilities regarding ROS crew manning. During ROS, the contractor:[2]

- Advises ROS crew members that duties aboard a ROS ship will differ significantly from standard mariner operations. Although officers in FOS have large administrative workloads, ROS status requires "hands-on" work that may not normally be associated with sea going (FOS) ships. ROS crews will be expected to comply with, but will not be limited to these general guidelines:
  - Tend to the general care and custody of the ship, including tending mooring arrangements and shore power cables, when fitted.
  - Develop familiarity with shipboard equipment and systems.

---

1 A vessel in ROS has a small/reduced crew to support the readiness of propulsion and other systems to activate the ship.

2 Military Sealift Command, Solicitation, Offer and Award, Solicitation No. N32205-16-R03000, issued June 10, 2016.

  - Operate, maintain and repair as necessary shipboard equipment and systems, including all associated documentation.
  - Perform all maintenance listed in the Shipboard Automated Maintenance Module (SAMM). . . . In this context maintenance includes regulatory body inspections, manufacturer's tests, preventative maintenance, corrective maintenance and repairs.
  - Comply with applicable Federal rules and regulations as may pertain to ships in ROS.
  - Be available and participate in inspections, audits and status checks as may be required by this contract and associated with ROS ships.
  - Inform/report on status of ships systems or equipment which may require additional repairs or upgrades.
- Provide onboard an ROS crew familiar with the ship's operating systems. The ROS crew shall sail with the ships upon transition to FOS. This does not preclude routine rotation of ROS crewmembers with qualified replacements.

Reduced Operating Status Crew Training Requirements and Challenges

The training requirements for ROS crews are similar to those of FOS crews. FOS crews must possess the required licensing, certification, and other required training prior to assignment to the vessel. It is the responsibility of the vessel-operating company that all mariners meet federal, state, and local training requirements.

Operating companies are required to ensure that each member of an ROS (and FOS) mariners crew are current and meet required federal, state, and local training requirements. Training requirements include but are not limited to fire prevention, helicopter firefighting, damage control, Standards of Training and Certification for Watchstanders (STCW) for their assigned rating, damage control, chemical/biological/radiological defense, small arms qualifications, and antiterrorism awareness training (annual), among other requirements.

Senior leaders voiced that the challenges experienced with manning the ships include sourcing crewmembers with the required MSC government training (e.g., Anti-Terrorism Force Protection, Chemical Biological Radiological Defense, small arms) and less with required federal ABS and/or USCG required training. This is usually caused by the low wage rates offered to the crews of surge ROS vessels when compared with other FOS ships. Mariners with required licenses can almost always be sourced through an operating company's labor unions, but crew with licenses and required training are more difficult to source as they normally sail on higher paying FOS contracts. Surge sealift officials continue to examine manpower demands needed to accomplish maintenance and repairs and have made marginal increases to vessel ROS crews accordingly. However, there is no data-based way to determine if the number of personnel assigned is sufficient.

How Reduced Operating Status Crews Differ from Full Operational Status Crews

The crews of FOS ships are on a different schedule than ROS crews. FOS vessels are operational, and crews can generally rotate on and off the ships, given the requisite training and certification requirements. Leaders with whom we engaged indicated that a nominal schedule for an FOS crewmember would be a four-month assignment on duty aboard the vessel followed by one and possibly two months off time. The FOS crewmember billet would then be filled by a replacement crewmember. FOS crews earn 1.5 days' vacation time for every month they work. During the duty aboard, an FOS crewmember is continuously working on their qualifications and time on watch to gain experience and certifications needed to support promotion to the next sequential grade (e.g., from 3rd mate to 2nd mate).

Replacement crews are those that relieve FOS crews when they rotate off the ship, and they must complete all required training prior to assignment to the vessel. There are courses that can only be completed on board the vessel, and a 30-day grace period is allowed for an FOS replacement crewmember to complete those training requirements once onboard.

In FOS, the key shipboard billets are the master, chief mate, chief engineer, first assistant engineer, and electronic officer. In ROS-5, the key shipboard billets are the senior officer in charge, chief engineer, and first assistant engineer. The majority of ROS crewmembers beyond the key billets are engineers; a cook and storekeeper fill out the staff. The number of crewmembers vastly differs between a FOS crew and ROS crew. For example, on a KOCAK class ship the number of FOS crewmembers is 30 while the number of ROS crew is 12.[3] ROS crews are also on a different and lower pay scale than FOS crews. FOS crews working on operational vessels are more highly compensated than ROS crews.

ROS crews gain very little watch standing experience that is needed to achieve their professional qualifications and certifications, as well as complete shipboard training requirements similar to that achieved by FOS crews. TAs can provide some underway experience for ROS crews in operating the ship and engineering plant, but TAs occur infrequently. Mission operations are very useful in qualifying the ROS crew. However, as noted, the number of ships performing missions are few, and underway days have been on a downward trend.

In view of these differences, there is little room for ROS crews' personnel growth and achievement of qualifications due to lack of underway time, and moreover, the lower pay scale is less than attractive to grow and sustain the force.

---

[3] U.S. Department of the Navy, U.S. Military Sealift Command, Military Sealift Command Program Office (PM-35), *ROCON Technical Manual*, Section 19, "Minimum Manning Requirements for WHEAT/MARTIN/KOCAK/PLESS/OBREGON Vessels," Washington, D.C., n.d.

How Reduced Operating Status Crews Are Managed

When we met with surge sealift leadership, we inquired about the status of ROS manning today; that is, are the surge sealift ships fully manned? USTRANSCOM reports that it is the ship's managers' responsibility to maintain the ships and to man the force. We learned that operating companies are required to provide crewmembers who are trained to contract requirements and are paid to provide the full ROS crew manning. If personnel are not provided, they are required to provide a credit for gapped billets in their invoices. Our discussion indicated that the operating companies rarely gap critical billets as vessel contractors need to fully staff the vessels to meet their best business interests. MSC does not track the gapping of billets by the operating companies. From our interviews, we learned that there are no contract data requirements lists in the contract that require the contractor to provide data on gapped billets, so the prevalence of gapped billets is not tracked. We were not provided personnel data that addressed the numbers of personnel and the qualifications or experience that could have provided greater transparency in assessing the sufficiency of ROS crews.

Are Mariners in Sufficient Supply to Meet Readiness Needs?

MARAD is tasked to determine if there is adequate manpower to support the full activation of surge sealift fleet. MARAD's assessment of the civilian U.S. merchant mariner pool shows that the number of civilian mariners available to crew government sealift ships when activated has declined over the past decade, and the current number of qualified and experienced mariners available may not be adequate in the near future.[4] The source of the manpower to man the surge sealift fleet are those mariners employed on U.S. flag vessels. The challenge is that the number of credentialed mariners has not increased proportionately with the increase in domestic trade; moreover, the number of U.S. flag vessels used for international trade has decreased, further reducing the mariner pool.

The Maritime Workforce Working Group[5] examined the aggregate number of qualified mariners required to activate the surge sealift fleet (composed of 61 MARAD and MSC surge sealift vessels) as well as the entire U.S. surge FOS fleet, which includes the commercially owned coastwise and oceangoing fleet, and MSC fleet. They estimated that 1,929 mariners are needed for the vessels to be fully crewed to meet current USCG requirements and mission accomplishment.

---

[4] U.S. Senate, Committee on Armed Services, Subcommittee on Seapower and Projection Forces, 114th Congress, 2nd Session, March 22, 2016 (testimony of Paul Jaenichen, on logistics and sealift force requirements).

[5] U.S. Department of Transportation, Maritime Administration, *Maritime Workforce Working Group Report*, Washington, D.C., September 29, 2017.

Congressional hearings on surge sealift manpower are well documented and provide an aggregate overview of manpower demand and the challenges with meeting it.

MARAD performed an exercise to determine the personnel readiness of the surge sealift fleet to meet full activation demands. The Strategic Sealift Command Post Exercise (CPX), done in March 2018, simulated the activation of 46 MARAD RRF vessels and the 15 MSC surge sealift ships. It was determined that a total of 1,876 billets (1,421 for MARAD and 455 for MSC) needed to be identified, by name, to support the activation. The exercise was geared to demonstrate the capability of the ship manager, operating company, and labor union processes to initially crew the RRF and MSC surge sealift vessels simultaneously.[6]

Findings from the CPX exercise showed that enough crewmembers could be identified to support the initial activation of the fleet. After the initial 180-day surge, when operators would be relieved, replacing steam engineers with qualified reliefs will be challenging. A second finding was that with sufficient lead time (greater than five days), the Navy reserve strategic sealift officer could be tasked with providing gapped billets, if required. The readiness exercise only considered the STCW readiness of the personnel assigned to ships to be activated. Finally, in the event of total activation, there may be insufficient numbers of personnel with small arms and antiterrorism qualifications, and training would be needed to support full employment.

While personnel assignments to ships were accounted for and evaluated by the MARAD team, the full qualifications of the crews were not verified. The availability and confirmation of the crews to report to and man the ships for an activation was inconclusive. A survey was sent via email to mariners identified by unions. Few of the total surge sealift fleet mariners (1,876) who were identified to report to a ship were contacted—78 total responses were received with 74 mariners (95 percent) responding that they would support the activation.

The shipmasters and engineers with whom we met indicated that for Operation Desert Storm there were enough ships and manpower to support the operation. However, for OIF/OEF there are enough ships to support the mission but not enough manpower. There is concern on the waterfront of the availability of manpower to support ship operations.

The conclusion of the CPX was that "there are enough mariners to conduct a full-scale breakout, but sustained operations past 180 days will be a crewing challenge." Since this was a synthetic exercise and mariners weren't directed to report to a vessel, the outcome could vary from mariners that report to a contingency billet.[7]

---

6 U.S. Department of Transportation, Maritime Administration, *Strategic Sealift Command Post Exercise Breakout 2018* (draft), Washington, D.C., March 20–29, 2018.

7 U.S. Department of Transportation, Maritime Administration, 2018, p. 13.

Summary

ROS crews' responsibilities are much more focused on maintenance and sustainment of the vessel than FOS crews. When FOS crews are assigned to a vessel, considerably more maintenance demands are accomplished with their addition. ROS crew mariners with required licenses and all required training are difficult to find, and it is challenging for ROS crews to gain all certification and training demands as compared with FOS crews; reduced or limited underway time plays a big role. Vessel leadership report that crew continuity is important to readiness and is a perishable resource, and the only way to pass knowledge on is to get underway with sufficient time at sea to get the crew their certifications. There is a national shortage of qualified personnel that directly affects the ability to man the surge sealift fleet—mariner manning may be sufficient for initial activation, but activation crews may have long waits for a replacement crew, especially for steam engineers.

CHAPTER FIVE

Management Best Practices

Some of the impetus for conducting this study came from the perception that MSC and MARAD were applying different management systems and getting substantially different results. Our review of the overall readiness requirement and the provision of personnel and material to meet these requirements suggests that there may be broader systemic issues that are not necessarily tied to one management construct over another. Issues include the following:

- Most ships are not routinely activated.
- The credibility of readiness assessments may be dubious.
- The fleet is aging and some of the vessels are unique, creating
  - parts obsolescence challenges
  - personnel expertise shortfalls (steam engineering for example).
- There is increasing demand for repair rather than preventive maintenance.
- The documentation of deferred maintenance and/or risk assumed from deferred maintenance is not well described.
- There are well-documented manning risks in FOS crews.
- There are shortfalls in ROS proficiency and manning levels that are not as well documented.

Changes That Could Be Applied Under Any Management Structure

Even if no changes were attempted in the alignment of maintenance responsibilities, the following items need to be undertaken. Perhaps most importantly, operational requirements need to be stated clearly and realistically reflect arrival times, given every other constraint in the delivery system. Sealift should not be required to be readier than what the rest of the delivery system can reasonably provide. In addition, the following other practices should be implemented:

- improved documentation of deferred maintenance and, in particular, the mission impact of deferral

- common access databases across the whole sealift enterprise
- streamlined processes for alterations to replace obsolete and difficult to maintain equipment
- improved oversight of ship management companies and repair contractors
- common standard for report of mission-limiting casualties
- investment in home port facilities—in particular, storm-protected harbor berths
- more extensive and more varied TAs that are more aligned with the vessel's operational employment.

These practices could be achieved by memoranda of understanding, by parallel policy action at MSC and MARAD, or by a more radical restructuring. However, the major objective of these improvements is to create a common view of readiness and to promote common practices across the stakeholders in strategic sealift to support surge sealift readiness.

The Case for a Single Readiness Manager

The most obvious objection to the dual management structure is that it does not appear to have originated from a clear decision with clear justification. With different structures, there are different perspectives, reporting standards, reporting systems, and two sometimes overlapping sustainment systems. It is inefficient on the face of it, and there does not seem to be a good reason for its existence in the first place. However, while there is no apparent good reason to keep the current structure, we also examine the reasons why changing it might be beneficial.

Provider Competition

Strategic sealift is part of a market, but it is a very constrained market in which there is relatively little demand for the service, other than that generated by MARAD and MSC. While there is certainly a worldwide shipping and ship sustainment industry, since MSC and MARAD are constrained to use U.S. providers, this market does not function with normal market mechanisms.

Among the things this market structure does is disadvantage the recipient of the service if it does not retain maximum market leverage. Although there is only one recipient of sealift, the U.S. government, USTRANSCOM, MARAD, and MSC essentially compete against each other for access to crews, maintenance resources, and supplies. They are not a big enough market together for suppliers to expand, but individually they do provide an option for suppliers who are displeased by some part of the arrangement with the other organization. For example, a supplier who finds the MSC-managed working capital fund construct unduly rigid in contract execution has the option of offering services to mission-funded MARAD. Conversely, if aspects of MARAD's oversight prove onerous, providers have the option of shifting to MSC.

The same market structure applies for labor. Merchant seamen have the option of employment with either organization and do not necessarily have an incentive to seek a long-term arrangement that ensures they will be immediately ready for activations. Reducing the choices that merchant seamen have for one organization or the other increases the ability of that organization to ensure availability. It does not necessarily increase demand, which could be a precondition for creating additional supply but instead creates something like a safe cartel in which limited numbers of sailors are available, but the employment is stable, the risks known, and the supply relatively secure.

Core Competencies

Each organization has core competencies. MARAD functions within the broader merchant marine environment, and its charter includes ensuring that a merchant fleet exists that is capable of supporting national defense needs. This includes regular interaction with the maritime industry, maintenance providers, unions, and all the other organizations that make up the broader community supporting sealift. Its charter requires that it understand the maritime industry thoroughly and be in a position to exploit economies of scale and best use of maritime resources.

MSC is the naval component of USTRANSCOM and has a variety of operational responsibilities, including the operational oversight of activated sealift. It is in a better position to appreciate military need than MARAD. Among other things, this enables it to accurately estimate the capability and capacity requirements for sealift. If MARAD is, by charter, best able to state the best combination of providers, MSC by mission is best able to state how much and what kind of capability is needed and at what point.

While there are certainly examples of organizations being both providers and consumers of readiness, having different organizations carrying out each function does allow greater focus on the things the organizations are chartered and equipped to carry out. In this particular case, dividing responsibilities between the organization that draws on the maritime industry to create and provide a ready force (MSC) and the organization that understands and sets the requirement (MARAD) appears to be a sensible approach.

CHAPTER SIX

Conclusions and Recommendations

As described in Chapter One, our research questions included the following:

- Is the sealift fleet ready to execute National Defense Strategy assigned missions?
- Are there sufficient ready ships and crews?
- How long would it take to reach the achieved readiness?
- What is the gap between the requirement and the provided capability?
- Do the organizational approaches yield different results?
- What is the relative cost of each approach?

While we used the above questions as our guidelines for research, our findings did not in every case conform exactly to the format we originally specified. By looking at limited duration activations, we were able to establish that the sealift fleet can carry out those missions, but we found reasons to doubt that the force is postured for a larger-scale activation. At a minimum, the systems intended to demonstrate readiness were found to be ineffective in showing the readiness of the force required to meet larger-scale activations. The ineffectiveness of this system also brings into question any assessment of time required or the seriousness of the gap. While we found that this lack of reliability is present in both the RRF and the surge sealift force, we did find that the divided management construct had an impact on readiness generation. The following findings and recommendations describe our general findings more specifically.

Strategic sealift remains important as a capability in current operations, having participated in operations ranging from humanitarian assistance to aviation aircraft testing. For limited purposes and durations, the sealift fleet has never failed to meet operational tasking.

However, when the focus is on planning for larger contingencies, the force's readiness is not clear. Most existing operational plans require some level of transportation by sealift, with more than one combatant commander likely to require services. Ability to meet these demands is problematic. There are several reasons, some applying across the whole sealift fleet and independent of particular management structures. Some others are either a result of differing management practices or would at least benefit from having a single readiness manager. We will consider first the conclusions and then recommendations for meeting the challenges.

Conclusions

For purposes of sealift, the required readiness outputs are (1) ability to meet real-world tasking at short notice and (2) readiness to support operations plans' TPFDD time lines. The informally stated but not codified requirement for readiness adopted by surge sealift leadership is 85 percent of ships in the sealift fleet available for mission tasking within five days of being activated.

Readiness inputs refer to the resources required for ships to meet the required operational outputs. These are generally funded by the U.S. Navy, although there are a variety of resource providers. These resources include

- scheduled maintenance availabilities in industrial facilities
- unscheduled repairs requiring specialized parts for installed equipment
- qualified personnel, specifically merchant mariners for ROS to conduct organizational maintenance and FOS crews to perform
- training to ensure crews are current in mariner skills
- operating days to support training and test equipment.

The current funding levels are expected to support readiness and allow the ships to activate in time to deliver cargo to a given area of operations and satisfy combatant commanders' critical warfighting requirements. No specific number or pacing operation has been specified.

MSC has reported over the last year that approximately 71 percent of its fleet is ready to meet the five-day readiness standard, while MARAD reports a higher number, an average of about 85 percent. Aspects of the processes currently in place should be modified to provide a more accurate picture of strategic sealift readiness and more accurate requirements against which to measure.

Recommendations

In the context of the above conclusions, we offer the following recommendations.

Operational requirements. Operational requirements need to be stated clearly and realistically reflect arrival times, given every other constraint in the delivery system. Current sealift readiness requirements are higher than what the rest of the delivery system can reasonably provide. Surge sealift should not be required to be readier than the rest of the system.

- **Recommendations:** Formally revise the readiness requirement for sealift. Align readiness requirements to TPFDD needs, and realistically account for potential delays from other components.

TA practices. USTRANSCOM tests the readiness of the fleet by conducting a test of vessels' ability to activate, through the no-notice TA process. When a vessel is directed to conduct a TA, all aspects of a vessel's activation must be completed within the five-day readiness period. While the number of vessels in the surge sealift fleet has been relatively constant, the number of TAs has varied over time. Some ships have done several TAs while others have had few or none. Moreover, TAs do not accurately reflect what a vessel and crew would need to do to accomplish their mission.

Instead, to support a mission, a surge sealift vessel should be activated, complete all necessary requirements to deploy, get underway, transit to the port of embarkation, and fully utilize all onboard equipment required to execute a mission. This type of activation test would provide many benefits, including a full test and run time for the vessel's engineering plant, experience for the crew, and a test of the full range of the platform's capabilities for which it is designed to operate. The trade-off is the cost in time and money, which competes with other resource needs.

- **Recommendation:** Revise the TA practice to regular activation of multiple units for multiple days underway to align with missions.

ROC/POE. Some ROC/POEs are outdated for surge sealift vessels. This requires that equipment no longer in use be maintained to meet ABS standards. Money, time, and resources are wasted. The process is currently time consuming and lengthy, sometimes to the point that requirements remain in place when it is clear that they are not applicable. A streamlined ROC/POE process could be beneficial.

- **Recommendation:** Review ROC/POE for ships to ensure relevant requirements.

Material readiness. A major unknown is the material condition of the many ships that have not recently activated. Even among those activated, the TAs have not been universally successful and for some ship classes notably unsuccessful.

Under agreement with ABS, USCG delegates responsibility for certifications to ABS. ABS conducts these assessments at intervals, and completion of these assessments is a condition for operations. Failure to complete these certifications can be a matter of teams not being available or substantial and unexpected casualties.

Only a few ships are completely current with their certifications; most have several certifications that are not current, and many ships in each fleet were all missing required certifications. Expired certification does suggest that there are issues with the management of ROS-5 ships that are not being captured in TA performance or elsewhere. Investments are needed in many areas of material readiness to better ensure ships are ready and capable of executing their missions—ranging from more directly addressing the issue of parts obsolescence to common databases to improved oversight of repair contractors.

- **Recommendations:** Improve cost-benefit analysis for repair versus replacement of equipment. Improve documentation of deferred maintenance and, in particular, the mission impact of deferral. Streamline processes for alterations to replace obsolete and difficult to maintain equipment; obsolescent parts add measurable cost and seriously affect readiness; obsolescent plants are increasingly difficult to man and maintain. Improve oversight of ship management companies and repair contractors. Use a common standard for report of mission-limiting casualties. Invest in home port facilities—in particular, storm-protected harbor berths.

Personnel readiness. ROS crews' responsibilities are much more focused on maintenance and sustainment of the vessel than FOS crews. ROS crew mariners with required licenses and all required training are difficult to find, and it is challenging for ROS crews to gain all certification and training demands as compared with FOS crews. Reduced and limited underway time plays a big role. In addition, the pay for ROS crews is lower than FOS crews. Vessel leadership report that crew continuity is important to readiness and is a perishable resource, and the only way to pass knowledge on is to get underway with sufficient time at sea so that crews can obtain their certifications. There is a national shortage of qualified personnel that directly affects the ability to man the surge sealift fleet. Mariner manning may be sufficient for initial activation, but activation crews may have long waits for a replacement crew, especially for steam engineers.

- **Recommendations:** Improve stability and capability of sealift crews. Conduct more frequent and longer underway periods. Review compensation packages.

Management structure. The dual management structure of the surge sealift fleet does not appear to have originated from a clear decision with clear justification. Maintaining two management structures results in different reporting methods and maintenance tracking systems, among other differences. MARAD and MSC essentially accomplish the same missions with the surge sealift fleet. MSC is the naval component of USTRANSCOM and has a variety of operational responsibilities, including the operational oversight of activated sealift. It is in a better position to appreciate their core competency of military need than MARAD. MSC by mission is best able to state how much and what kind of capability is needed and at what point. Realigning responsibilities along these lines would result in a more efficient and effective management structure.

- **Recommendations:** Refocus MSC attention to USTRANSCOM component issues and away from day-to-day management of equipment and personnel. Align the man, train, maintain, and equip functions of MSC vessels to MARAD.

These recommendations are in some cases applicable regardless of management structure. More frequent and more varied TAs, more underway time for FOS crews, better documentation of material condition, and improved requirements review processes could all be implemented with or without a common management structure. However, the divided management structure has no clear justification or purpose and is in some respects detrimental to improvement. While changing the management structure to focus MSC on operational employment and MARAD on readiness generation does not really involve adding or subtracting staff, it would involve organizational change. Organizational change is often disruptive and should not be undertaken without good reason. The concerns surrounding strategic sealift's overall readiness appear to provide good reason.

Bibliography

Buzby, M. H., *Activation of Surge Strategic Sealift Ships*, U.S. Department of the Navy, Washington, D.C., COMSCINST 4626.2, 2012a.

———, *The U.S. Navy's Military Sealift Command Standard Operating Manual*, U.S. Department of the Navy, Washington, D.C., COMSCINST 3121.9C, 2012b.

Cahill, W. H., *Memorandum: Policy Directive 12-02*, U.S. Department of Transportation, Maritime Administration, Washington, D.C., 2012.

Crenshaw, L., Mortimer Downey III, Beverly Godwin, William Kenwell, and Marvin Phaup, *Maritime Administration: Defining Its Mission, Aligning Its Programs, and Meeting Its Objectives*, Washington, D.C.: National Academy of Public Administration, 2017. As of January 29, 2019:
https://www.napawash.org/uploads/Academy_Studies/11.27.17_MARAD_FINAL_Report.pdf

Gortney, W. E., "Force Readiness Reporting," Joint Chiefs of Staff, Washington, D.C., CJCSI 3401.02B, 2011. As of January 29, 2019:
https://www.jcs.mil/LinkClick.aspx?fileticket=nUmL4xAjQ3Y%3d&tabid=19767&portalid=36&mid=46626

Management Systems Consultants, *MARAD NS5 Maintenance Protocol*, Sugar Land, Tex., 2016.

"Memorandum of Agreement Between Department of Defense and Department of Transportation," 2008. As of April 10, 2019:
https: //www.gps.gov/policy/docs/2008-dod-dot-agreement.pdf

Military Sealift Command Public Affairs, "Sealift for Operation Iraqi Freedom Is No Small Thing," Washington, D.C., March 2, 2004.

OPNAV Instruction 3501.199C, *Required Operational Capabilities (ROC) and Projected Operational Capabilities (POE) for Strategic Sealift Ships to Include the T-AKR Fast Sealift Ships (FSS), Large, Medium-Speed RO/RO (LMSR), Aviation Support Ships (T-AVB), Auxiliary Crane Ships (T-ACS), and Ready Reserve Force (RRF) Dry Cargo Ships*, U.S. Department of the Navy, Washington, D.C., 2007.

U.S. Coast Guard Maritime Information Exchange, "Port State Information Exchange," webpage, January 28, 2019. As of January 29, 2019:
https://cgmix.uscg.mil/psix/

U.S. Department of Defense, Office of Inspector General, *Military Sealift Command Oversight of Excess Spare-Parts Inventory and Purchases for Sealift Program Roll-On/Roll-Off Ships Needs Improvement*, Alexandria, Va., DODIG-2014-106, 2014.

U.S. Department of the Navy, "COMSC INSTRUCTION 4700.15D," Washington, D.C., COMSCINST 4700.15D, n.d.

———, Department of the Navy Fiscal Year (FY) 2018 Budget Estimates, Justification of Estimates, May 2017, National Defense Sealift Fund, Washington, D.C., n.d.

———, U.S. Military Sealift Command, Military Sealift Command Program Office (PM-35), *ROCON Technical Manual*, Section 19, "Minimum Manning Requirements for WHEAT/MARTIN/KOCAK/PLESS/OBREGON Vessels," Washington, D.C., n.d.

———, Military Sealift Command, Solicitation, Offer and Award, Solicitation No. N32205-16-R03000, June 10, 2016.

———, Naval Center for Cost Analysis, *Documentation for NCCA's Assessment of MSC and MARAD Sealift Cost*, Washington, D.C., 2018.

U.S. Department of Transportation, Maritime Administration, homepage, n.d. As of January 29, 2019:
https://www.maritime.dot.gov/

———, Maritime Administration, *Ready Reserve Force Logistics Management Manual*, Washington, D.C., 2013.

———, Maritime Administration, *Maritime Workforce Working Group Report*, Washington, D.C., September 29, 2017.

———, Maritime Administration, *Strategic Sealift Command Post Exercise Breakout 2018* (draft), Washington, D.C., March 20–29, 2018.

———, Office of the Secretary of Transportation, Office of Inspector General, *Weaknesses in MARAD's Management Controls for Risk Mitigation, Workforce Development, and Program Implementation Hinder the Agency's Ability to Meet Its Mission: Maritime Administration*, ST-2016-011, December 10, 2015.

U.S. General Accounting Office, *Ready Reserve Force: Ship Readiness Has Improved, but Other Concerns Remain*, Washington, D.C., GAO/NSID-95-24, 1994a. As of January 29, 2019:
https://www.gao.gov/products/NSIAD-95-24

———, *U.S. Navy/Military Sealift Command: Weak Contract Administration Led to Unsafe and Poorly Maintained Ships*, Washington, D.C., OSI-94-27, 1994b. As of January 29, 2019:
https://www.gao.gov/products/OSI-94-27

———, *Military Sealift Command: Weak Controls and Management of Contractor-Operated Ships*, Washington, D.C., NSIAD-96-41, 1995. As of January 29, 2019:
https://www.gao.gov/products/NSIAD-96-41

U.S. Government Accountability Office, *Navy Readiness: Actions Needed to Maintain Viable Surge Sealift and Combat Logistics Fleets*, Washington, D.C., GAO-17-503, 2017. As of January 29, 2019:
https://www.gao.gov/products/GAO-17-503

U.S. Senate, Committee on Armed Services, Subcommittee on Seapower and Projection Forces, 114th Congress, 2nd Session, March 22, 2016 (testimony of Paul Jaenichen, on logistics and sealift force requirements).

U.S. Transportation Command, Memorandum, CCJ5/4, May 19, 2015.

World Maritime News Staff, "US Congress Passes El Faro Maritime Safety Act," *World Maritime News*, September 28, 2018. As of January 29, 2019:
https://worldmaritimenews.com/archives/261510/us-congress-passes-el-faro-maritime-safety-act/